HOW TO CATCH A
LOBSTER
IN DOWN
EAST
MAINE

HOW TO CATCH A LOBSTER IN DOWN EAST MAINE

CHRISTINA LEMIEUX ORAGANO

Charleston   London

THE
History
PRESS

Published by The History Press
Charleston, SC 29403
www.historypress.net

Unless otherwise noted, all images are from the author's collection.

First published 2012

Manufactured in the United States

ISBN 978.1.60949.602.9

Library of Congress CIP data applied for.

Notice: The information in this book is true and complete to the best of our knowledge. It is offered without guarantee on the part of the author or The History Press. The author and The History Press disclaim all liability in connection with the use of this book.

To all the lobster-fishing families of Down East Maine,
especially my own.

CONTENTS

ACKNOWLEDGEMENTS

As with any big project, this book was only possible due to the help and support of many people. I owe special thanks to my husband, Anthony, for being open to me taking on this project, especially knowing how much time it would involve. The opportunity to write this book came to me just months after Anthony and I had our first child, when our lives were already quite hectic. Anthony willingly gave up many of his own pursuits to help out with the book—from managing much of the photography to giving extra care to our daughter, Anya, so I could focus on research and writing.

I am grateful to the Downeast Lobstermen's Association (DELA) for helping me enrich this book with feedback from Down East fishermen. Special thanks go to Sheila Dassatt, executive director of the DELA, for being so supportive and helping me reach out to all the DELA fishermen. I am also grateful to the many fishermen and women who took time to fill out and return the surveys I sent. I know it was often the wives who oversaw the completion and mailing of the surveys, and I appreciate your efforts. Your perspectives and comments made me smile and laugh out loud at times.

Thanks as well go to all the people who shared photos for the book. Sadly, I was not able to include some photography due to resolution issues, but I appreciate all of your efforts. Special thanks go to Billy Kitchen, who was open to me using some of his beautiful photography.

Thanks to the Maine Lobstermen's Association and the Maine Department of Marine Resources for helping me underpin my writing with facts and examples. Thanks as well to Pauline Cates, Patricia Pashazadeh and Natalie Gingerich for taking time from their busy lives to read through my writing and give me valuable feedback before I sent the manuscript off to the publisher. Thanks to Whitney Tarella, my editor at The History Press, for being patient and supportive through all the struggles I faced during the writing process.

Finally, I owe a special thanks to my brother, my father and my mother. My brother, Nick, has always been an inspiration to me. He is the reason I started lobster fishing in the first place. Even at the age of ten, my desire to emulate my brother was so strong that I declared to my father, "I want to go sternman on the boat like Nick does." Even though no other girls in my town were sternmen at the time and the task involved getting up at 5:00 a.m., I was glad to work on the boat if it meant being a bit more like Nick.

Thanks to my father for supporting me with all my pursuits in life—from taking on a girl for a sternman to helping me write this book. My father has shared considerable time and knowledge with me to help ensure this book is accurate, informative and interesting. Working with him on various chapters—from boat design to the Down East tidal system—has been a joy as well as a wonderful reminder to me of his incredible wealth of knowledge.

And lastly, a special thanks to my mother. I wish she could be here to see this in print. As with all of my undertakings, Mum helped me in so many ways with this book. She helped me print, post and keep track of the surveys. She helped me source many of the photos. She also spent considerable time taking care of Anya and helping out around the house so that I could focus on writing. Above all, she always believed I could accomplish whatever I set my mind to—be it raising a family, advancing my career or writing this book.

INTRODUCTION

MY LOBSTER-FISHING HERITAGE

If you follow the contours of the Maine coast, past the towns of Kennebunkport, Portland and Searsport, past Bar Harbor, Beals Island and Machiasport, you'll finally come upon a Lilliputian harbor. The harbor is fed from one end by a little river, conveniently called Little River, and protected at the other by a small island, upon which sits a working lighthouse. A smattering of boats rests in the harbor, and the surrounding land plays host to approximately one hundred modest houses. Welcome to the village of Cutler.

Cutler Village currently has no convenience store, restaurant, gas station or gift shop. In fact, people traveling to Cutler for the first time often miss the town. Only after passing through the village and coming upon the LEAVING CUTLER sign do they realize they've gone too far. As described in a Massachusetts Institute of Technology (MIT) report, Cutler's town center "exists as little more than a curve in the road linked to a small cove, with docks for local lobster crews and a small storage warehouse for their gear."[1]

This is the town in which I was born and raised. To those passing through, Cutler can appear to be just a sleepy village, the kind that makes for great poetry and postcards. People who pause and spend some time, however, discover a bustling, working waterfront and community of individuals determined to wrestle a living from the beautiful, yet unforgiving, ocean. For the last century or so, that living has come primarily from lobster fishing.

Like many towns along the Maine coast, Cutler harbors lobstering families who can trace back their ancestry four or more generations. This is the case with my family. My great-grandfather got into the lobster business sometime in the early 1900s, operating a lobster smack. He later ran a lobster business, buying and selling lobsters in our town and traveling to nearby harbors and inlets to purchase lobsters from fishermen who didn't have an established dealer. My grandfather took over where my great-grandfather left off, operating a lobster dealership while he and my grandmother raised two daughters. He also fished for lobster along the Cutler coast on and off throughout his life. Though my father was born in central Maine and is therefore technically "from away," shortly after marrying my mother, he and she settled in Cutler. By the time I was born, my father had begun carving out a career as a full-time lobster fisherman. This career has, again, been handed down to my brother, who is currently running a successful lobster-fishing business while raising two sons, sons who may well become fishermen when they get older.

Not only do I come from a long lineage of lobstering, but I have also played an active role in the fishing industry myself. Around the age of eight, I started painting my father's lobster buoys and helping him repair his traps. At the age of ten, I began working as a sternman on his boat. Day after day, summer after summer, I would rise by 5:00 a.m. and spend my day on the ocean, stuffing bait bags and banding lobsters. Going sternman remained my summer job until I graduated from college. Through the years, I estimate I have easily spent five thousand hours on a lobster boat.

Even after finishing university, it was my lobster-fishing heritage that helped me secure my future career in advertising. Months of interviews and a *summa cum laude* degree from Colby College didn't seem to be enough to help me land a position as an account executive at any of the leading ad firms. It was only after a serendipitous meeting between my father and a visitor to Cutler harbor that doors started opening. As the story goes, a man had navigated his pleasure craft into our modest harbor in search of someone to fix his problematic engine. The visitor was eventually referred to my father, who, in addition to lobster fishing, is also a skilled engineer. After fixing the engine, my father and the man engaged in a lengthy conversation, which eventually led to stories of my lobster-fishing past and my present hopes to work in advertising. At the time, my father had no idea who the man was. As it turns out, the man was chairman and CEO of one of the world's leading advertising agencies. Needless to say, the next week I had a job.

My brother and me "playing" lobster fishing in our front yard in the early 1980s. The miniature skiff, wharf and trap were built by my father and grandfather.

My brother and me lobster fishing with my father in the late 1980s. *Courtesy of Peter Ralston.*

My father and me lobster fishing in the summer of 1996. *Courtesy of Eric Piper, www.facebook. com/PIPERGRAPHIC.*

My career in advertising has taken me on many journeys. Currently, I live and work in London as a strategy director for a digital advertising agency. Yet all roads seem to lead me back to my lobster-fishing heritage. When blogging became a popular pastime in the noughties, I began writing a blog about lobster fishing. Several years later, the blog was discovered by The History Press publishing company, which asked me to write this book.

The thought of writing a book about lobster fishing was, for me, a daunting one. How could I author a story about Maine lobstering while living in a foreign country? Did I even have the right to write such a book, given that I was now technically "from away"? On top of all this, where would I find the time? I was already fully employed and had recently given birth to my first child. Yet these worries were surpassed by my strong belief that this was a story worth telling.

I chose to focus my book on lobstering in Down East Maine, as it is where I am from and because lobster fishing along this stretch of Maine coast is a way of life more than an occupation. In an attempt to paint an accurate picture of the lobstering industry Down East, I have supplemented my personal experiences and viewpoints with some qualitative and quantitative research. For example, data from Maine's Department of Marine Resources (DMR) has helped me understand and describe key demographics, such as the relative size of lobster-fishing communities and the conventions used when naming boats. A questionnaire to Down East fishermen has helped me get closer to the hearts and minds of many of these rugged individualists.

One particular detail I wish to acknowledge up front is that most references in this book are in the masculine tense. "Fisherman," "sternman" and "he" or "his" are used to describe most fishing scenarios. My choice of language is reflective of the fact that only a small handful of full-time Down East "lobstermen" are women. Many of these women, including myself, are happy to be referred to as a "sternman" or "fisherman." It's almost a rite of passage. Yet all women in the lobster industry deserve special recognition for making their way in what is a very male-dominated occupation. I hope this book will serve as an anthem to all the lobstermen and women in Down East Maine, acknowledging their hard work and dedication to preserving a beautiful way of life for future generations.

WHERE THE HECK IS DOWN EAST MAINE?

In the most limited sense, Down East refers to the stretch of Maine coast from Penobscot Bay to the Canadian border. Within Penobscot Bay, the town of Searsport is often earmarked as the start of "Down East." To an outsider, "Down East" is a contradiction of terms, given that the entire Maine coast runs a northerly course. But "Down East" is actually a sailing term that refers to direction, not geographical location. Earlier generations used to travel to Maine from Boston via sailboats, leveraging the prevailing southwest winds to fill their sails. Given that the wind was at their backs, these vessels sailed "downwind" to reach the easternmost points of Maine, hence the term "Down East."

Down East Maine encompasses both Hancock and Washington Counties and includes bustling fishing villages like Bucksport, Searsport, Stonington, Northeast and Southwest Harbor, Winter Harbor, Prospect Harbor, Jonesport, Eastport and my hometown of Cutler. Yet even native Down Easters like myself debate what towns, islands and inlets compose Down East Maine. At times, Down East Maine is jokingly referred to as any point along the coast, east of the speaker.

As part of my research for this book, I engaged with fishermen from the Downeast Lobstermen's Association (DELA). In addition to all of Washington and Hancock Counties, DELA fishermen hail from some parts

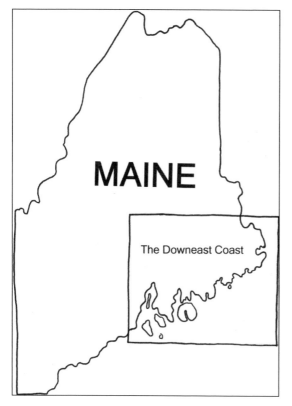

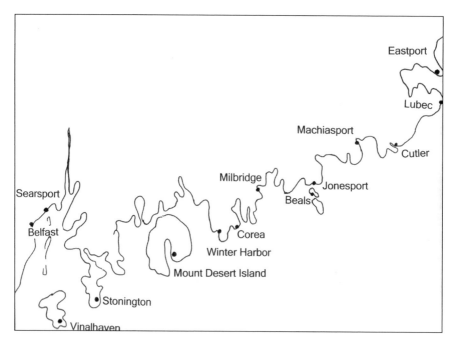

Left: The state of Maine and Down East region. *Sketch by the author.*

Below: The Down East coast. *Sketch by the author.*

of Knox County, such as Vinalhaven and North Haven. For the purposes of this book, I have stuck rather closely to the DELA's view of Down East Maine and included the towns, islands and inlets east of, and including, Bucksport in my definition. I have also included the islands of Matinicus, Vinalhaven, North Haven and Mount Desert. The qualitative and quantitative research I've conducted for this book is based on these above-mentioned localities and hopefully provides an accurate and informed picture of Down East Maine lobster fishermen.

THE FISHERMEN AND FISHING VILLAGES OF DOWN EAST MAINE

Despite their long-standing poverty, Down East Mainers live on what is perhaps the final stretch of undersettled and underdeveloped coastline on the entire US eastern seaboard.
—The Lobster Coast

Of the 4,300 active fishermen in Maine, more than half are based in Down East Maine. In fact, approximately 2,890 Down Easters hold some form of lobster-fishing license. These fishermen come in all shapes, sizes and ages. While the average age of a Down East lobster fisherman is forty-three years old, there are fishermen as young as eight and as old as ninety-four.[2]

In these parts, lobster fishing is not just an occupation—it's a way of life and a family tradition. Children grow up "playing" lobster fishing in old skiffs parked in their parents' front yards. Boys, and some girls, begin fishing sternman with their fathers as young as the age of eight. By their early teens, many boys are fishing a small gang of their own traps from an outboard boat. Before they graduate high school, these boys have often graduated to full-scale lobster boats.

Lobster fishing is not a skill learned in school; rather, it is a vocation handed down from generation to generation, as son works alongside his father. Of the fishermen I surveyed for this book, 85 percent are part of a generation of lobster fishermen. Most of these fishermen had a father and a grandfather who worked as lobster fishermen and taught them the craft; 40 percent also had a lobster-fishing great-grandfather. One of my survey respondents came from five generations of lobster fishermen.

Handing on one's skills to the next generation continues to be ubiquitous with the lobster-fishing industry today; 80 percent of Down East lobstermen

Left: Lobsterman Jasper "Cappie" Cates, who fished in Cutler most of his life and passed the legacy of fishing on to his sons, grandchildren and great-grandchildren. *Courtesy of Wanda Cates.*

Below: Brian Cates, son of Jasper Cates, lobster fishing with his two sons, Jeremy and Joshua Cates. *Courtesy of Wanda Cates.*

The legacy continues. Jasper Cates's great-grandson, Lucas Cates, learning about lobster fishing from his father, Josh. *Courtesy of Laurie Cates.*

have their children help them on the boat. Once those children grow up, over half become full-time fishermen themselves.

These multigenerational fishermen operate out of ninety-two harbors, islands and inlets along the Down East coast. The biggest of those fishing villages are the islands of Vinalhaven and Beals, where at least 25 percent of the population hold a lobster-fishing license and upward of 75 percent of the community are directly dependent on the industry. Deer Isle, Stonington, Jonesport and Mount Desert Island also have some of the largest gangs of lobster fishermen in Down East Maine. While some smaller harbors and inlets house only a handful of fishermen, for most communities in Down East Maine, lobster fishing is deeply woven into the fabric of everyday life.[3]

As an individual, the lobster fisherman holds a certain allure for the rest of America. In some ways, he is the last of the rugged individualists. As summed up brilliantly by James M. Acheson:[4]

> *He is his own boss and his own man, willing to defend his independence with violence if necessary. His daily activities are dictated by weather and the turn of the seasons, rather than by the office clock, governmental bureaucracy, or society's expectations. Fishermen tend to present themselves*

to tourists as men who earn their living from a relentless and icy sea with nothing but their skill, courage and tenacity. If sophisticated urbanites chuckle at the rustics on the Maine docks, they do so with a tinge of envy, for the lobster fishermen embodies many of our most cherished virtues. He is, along with the farmer and rancher, the quintessential American.

While a twinge romanticized, this quote from *The Lobster Gangs of Maine* is a largely accurate portrayal of Down East Maine lobstermen, past and present. Lobstering, unlike almost every other American fishery, has managed to survive as an artisanal enterprise well into the twenty-first century. Though the marketing and distribution of lobster operate on a colossal, corporate scale, the actual harvesting of lobster is still undertaken by a dispersed fleet of boats owned and operated by independent, self-employed individuals.[5]

This independence is greatly prized and fiercely defended by the fishermen. When asked what they most value about their job, almost 80 percent of Down East fishermen spontaneously respond either "independence," "being my own boss" or "freedom." What fishermen like least and worry about most when it comes to their jobs are government regulations. While it's doubtful this response is unique to just the fishermen of Down East Maine, what is unique is the reliance Down East communities have on the lobster-fishing industry. A Gulf of Maine Research Institute report determined that Down East communities remain the most fishery-dependent communities of all of New England. If fishing were to cease in Down East Maine, there would be, on average, two and a half fishermen available to fill any single similar occupation.[6]

The dependence on fishing in Down East Maine is due largely to how isolated this region of Maine is to the rest of the state and the country. Only U.S. Route 1 and Maine Routes 9 and 6 inland navigate through Washington County. Some of the residents in these villages rarely leave the area and reflect a cultural uniqueness born of their dependence on the natural resources and minimal interaction with the outside world. For example, the MIT report described talking to Cutler locals as "talking about lobsters, about independent thinking and action [and about] self-reliance."[7]

Up until recently, most Americans were so accustomed to the image of independent lobstermen like those from Cutler that they didn't question how the industry had escaped corporate consolidation. Only recently, as the Maine lobster fishery has continued to defy the doom and gloom warnings from government and scientists, has a credit been given to how the fishermen themselves responsibly maintain their precious resource. In fact, in 2009, Elinor Ostrom won a Nobel Prize in economics in part by studying how common resources—including lobster fisheries in Maine—are successfully managed by communities. In particular, and to the delight of many fishermen, one of her key findings was that self-governance often worked much better than an ill-informed government taking over and imposing occasionally clumsy and often ineffective rules.[8]

The methods by which Maine lobstermen have long managed their fishery have today become a textbook example of how communities successfully protect and defend the resources upon which they rely. The common anthem of the lobstermen is, "Let us manage this ourselves, we know what we're doing." The current abundance of lobsters up and down the coast of Maine is testimony that they might just know what they're talking about.[9]

A BRIEF HISTORY OF LOBSTER FISHING IN MAINE

THE 1600S POVERTY FOOD

Lobsters have been part of the New England diet since the colonial days. The first official reported lobster catch came from a group of English settlers in 1605. In an account of their voyage to Maine, one of the settlers wrote: "And towards night we drew with a small net of twenty fathoms very nigh the shore; we got about thirty very good and great lobsters."[10]

Captain John Smith of Virginia, who visited Monhegan and midcoast Maine in the early 1600s, described the bounty of lobsters along the shores. The lobsters were alleged to be not just plentiful but gigantic as well, with some measuring up to five feet in length.[11]

In those days, lobsters were so plentiful they would wash up on beaches and could be collected simply by wading along the shores at low tide. Native Americans gathered lobsters to fertilize their fields, and European colonists served them to prisoners and indentured servants. Considered "poverty food," lobster was fed to servants so frequently that there are stories of discontent and rebellions. One group of indentured servants even took their owners to court, winning a judgment that they would not be served lobster more than three times a week.[12]

The Late *1700*s The Lobster Smack

Real commercial exploitation of the lobster did not begin until the advent of the lobster smack in the late 1700s. Unlike most other kinds of fish, lobsters must be transported live. The development of the lobster smack, a sailing vessel with circulating seawater tanks in its hold, made it possible to ship lobsters cost-effectively to the New York and Boston markets. The first smacks were developed in the late 1700s and began operating in Maine in the early 1820s. The arrival of the smack boat to the isolated coastal villages of Maine was an important event. The smack skipper would share news and gossip with the fishermen and sell them fishing and household supplies on credit. Lobster smacks operated well into the 1900s in Down East Maine.[13]

The Mid-*1800*s Lobster Canning

The largest catalyst for the lobster industry came from the tip of Down East Maine when, in 1842, the town of Eastport started canning lobster. It was the first time the world had seen canned lobster and the second time Americans had ever canned anything at all. The canning process overcame the hurdle of getting getting fit-to-eat lobster from point A to point B cost effectively. In 1843, a one-pound can (meat from three and a half pounds of live lobster) sold for five cents. Soon, these cans were flying off the shelves. By the late 1870s, there were twenty-three canneries between Portland and Eastport, supported by lobster-fishing fleets in most villages along the coast of Maine. These fishing fleets were now catching their lobsters in lath pots, which had replaced hoop nets in the 1850s. A typical lobsterman in 1860 would tend twenty-five to fifty pots from an oar-powered dory. An average catch in those days was an astounding seven four- to six-pound lobsters per pot.[14]

In order to meet the nation's demand for lobster meat, the canneries began sourcing smaller and smaller lobsters from the fishermen. Soon half-pound "snappers" were being used to stuff their tins, leading to a decline in the lobster stock. This stock decline led to Maine's first lobster-fishing regulations. In 1872, a law was passed prohibiting the taking of females bearing eggs. Two years later, another law was put in place outlawing the harvesting of lobsters under 10.5 inches.[15]

The minimum-size law helped seal the fate of the lobster canners. By 1885, most of Maine's canneries were out of business. Today, there are few remaining lobster canneries in Maine. One of these canneries is located just

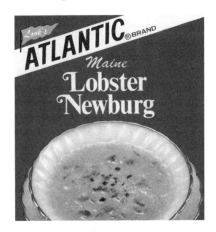

Down East Maine canned lobster products from the early 1900s. *Courtesy of Cynthia Fisher.*

outside of Cutler. It has been in operation since 1917, serving lobster meat, bisque, newburg and more to America, Europe and the Far East. Through the years, the cannery has also marketed such novelty products as lobster pet food and "swam," the seafood equivalent of spam.[16]

THE LATE *1800*s THE LIVE LOBSTER MARKET

Three events in the late 1800s helped set the stage for today's vibrant live lobster market: the laying of railroad tracks in Maine, the invention of the lobster pound and the first influxes of tourists to Vacationland.

In the 1870s, railroad tracks began linking western Maine with the metropolises of Boston, New York and beyond, allowing lobstermen to pack their lobsters in ice and ship them live throughout the country. One story attributes the newspaper tycoon William Randolph Hearst with the first order of live lobster for a dinner party in Colorado. Shortly thereafter, the first lobster pound appeared on Vinalhaven. Working on the same principle as smack boats, the pound kept lobsters in tanks with water passing freely through them. Using the pound, lobster dealers were able to store lobsters for extended periods of time, supplying live lobsters to the market in line with consumer demand.[17]

By the late 1800s as well, Maine was fast becoming a popular vacation spot for well-heeled Americans in search of picturesque landscapes and fresh, healing ocean air. By the early 1870s, Bar Harbor alone was home to fifteen large summer hotels, and summer cottages began dotting the coast. The arrival of these summer folk provided a critical boost to Maine's inshore fisheries. Their appetite for fresh seafood expanded existing fresh seafood markets and fueled new ones as well. Clam bakes and lobster boils soon became fixtures of the Maine vacation experience.[18]

The Early 1900s Low Landings

In the early 1900s, lobster landings were very low, fueled in part due to overfishing. The stock market crash of 1929 and the Great Depression of the 1930s dealt the fishery another blow, depressing demand for what was now considered a luxury item. Fishermen often had to find other ways to supplement their income in these days, including rum-running along the Maine coast during Prohibition. The design of lobster boats advanced significantly during this period, as bootlegging lobstermen invested in vessels that could outrun government patrol boats. One fisherman even rigged his boat with a secret underwater exhaust so on foggy nights he could silently swish past the revenue cutters on his way to collect alcohol.[19]

The Late 1900s to the Present The Golden Age of Lobster Fishing

Starting in the 1950s, Maine lobster landings began to steadily increase. Mounting bureaucratic pressure to regulate the industry led to the formation of the Maine Lobstermen's Association (MLA) in 1954. The MLA and subsequent Downeast Lobsterman's Association (DELA) have helped give a united voice to the geographically dispersed lobstermen of Maine. These organizations have helped the fishermen fight for common-sense conservation solutions, such as the maximum gauge and V-notching, to ensure that resource management is balanced with economic survival. Time and time again, Maine lobstermen have demonstrated that they know how to conserve the resource and keep the Maine lobster fishery healthy.

Not only has the Maine lobster catch proven remarkably steady since the 1950s, in the last several decades it's practically exploded. In the last fifteen years, lobster landings have increased by over 400 percent, from 20.1 million pounds of lobster in 1985 to almost 94.7 million pounds in 2010. Lobster now represents 80 percent of Maine's commercial fishing

A Brief History of Lobster Fishing in Maine

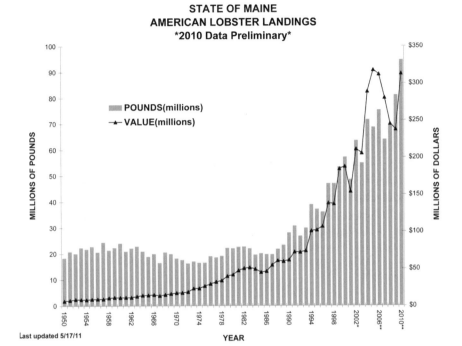

STATE OF MAINE
AMERICAN LOBSTER LANDINGS
2010 Data Preliminary

Last updated 5/17/11

Above: The dramatic rise in Maine lobster landings since the 1950s. *Courtesy of the Maine Department of Marine Resources (DMR).*

Right: Three generations of Cutler lobster fishermen working side by side. *Courtesy of Laurie Cates.*

income. Where overfishing has led to the decline of other Maine fisheries such as cod, haddock and halibut, smart conservation measures have helped preserve the lobster stock. For every lobster that comes onto the market, approximately three are sent back into the water for conservation. Baited lobster traps on the ocean floor help this brood stock thrive by supplying it with a steady food supply.[20]

It's surprising to think that while recent annual harvests have topped $300 million, modern Maine lobster fishing still operates much like it did in the older days. While the technology has advanced greatly, fishermen still own and operate their own boats and the industry remains primarily a family affair, with techniques and territories passed from one generation to the next. Maine, and especially Down East, is a close-knit community of fishermen who take care of and watch out for their own.[21]

LOBSTER-FISHING LINGO

Like people in other specialist fields of work, lobster fishermen use unique jargon when speaking about their profession. Throughout the book, I will often be using the language leveraged by lobster fishermen. Below is a list of the more common lingo.

gear: Fishermen often refer to their traps, line and buoys as their gear. In the spring, a fisherman will "set off gear," and he may speak of going out to "tend some gear" during the fishing season.

molesting gear: When one fisherman tampers with the traps (gear) of another. This can range from hauling another fisherman's traps and taking the lobsters to cutting another fisherman's trap line, causing him to lose the trap.

gang: Fishermen will often refer to the entire group of traps they fish as their gang. In Down East Maine, the largest gang of traps a person can fish is eight hundred.

sternman: A lobsterman's assistant, who works alongside the fisherman on his boat. He or she will bait up bait bags, help pick the traps, band lobsters and scrub the decks in return for a percentage of the catch. The sternman typically does not haul traps or pilot the boat.

going out to traps: When a fisherman heads out to haul his traps and collect the lobsters, he'll say he's "going out to traps."

tending a trap: The process of hauling up one's traps, collecting the lobsters and changing the bait bag is often referred to as "tending the traps."

setting traps and letting traps set: When fishermen put out their traps in the spring, it is referred to as "setting traps." The period between trap hauls is often referred to as "setting" as well. In order to catch more lobsters, a fisherman will let the trap set a bit longer between hauls. Other times, fishermen will say they're "letting their traps soak."

getting "bit" by a lobster: If a lobster gets ahold of a fisherman with its claw, the fisherman will often say he's been bit by the lobster. Though lobster claws have no teeth and pinching is a more accurate way to describe the experience, the word biting is often used.

baiting up: The process of getting bait on the boat or baiting up bait bags.

warp: A length of trap line. The line is measured in fathoms, with a fathom being six feet in length.

tangle or snarl: When the lines of two traps get knotted together, the mess will be referred to as a "tangle" or "snarl."

shorts or juveniles: Undersized lobsters that must be returned to the sea as a conservation measure.

eggers: Female lobsters carrying eggs, which must be returned to the sea as a conservation measure. Egg-bearing lobsters are also often referred to as berried females.

V-notcher: A specially marked female lobster, which must be returned to the sea as a conservation measure.

bull: A large, usually oversized male lobster.

keepers: Legal-sized lobsters with no other conservation restrictions, which can be taken home and sold by the fisherman. Though a fisherman will catch a variety of lobsters in his trap, he'll only count the keepers. For example, a trap may come up with three shorts, an egger and one legal lobster. If asked how the trap fished, the fisherman will say he caught one lobster.

shedders: Soft-shell lobsters.

cull: Any lobster that is not a marketable two-claw lobster. The lobster may be completely missing one claw or in the process of regrowing a claw.

dummy or pistol: A lobster with no claws.

boat price: The price per pound of lobster offered to fishermen by lobster dealers. This price fluctuates based on supply and demand.

flood and ebb tides: The tides as they build toward high and low water.

slack: The period on either side of high or low tide where the water is completely unstressed and free of movement. Down East fishermen often do their hauling during the tide slacks, as this is the only time lobster buoys aren't dragged under by the pull of the tides. Fishermen often speak of "hauling a slack" or "going out for a slack."

watching: A term to describe when a fisherman's buoys are visible on the surface of the water due to slack tides.

springing tides: The tides as they build toward their highest point of the month, which happens around the full moon.

onshore and offshore winds: Onshore winds are winds from the south. Offshore winds are winds from the north.

knots: A term used to describe the traveling speed of the lobster boat. Each knot equals 1.15 miles per hour. The fastest lobster boats can run fifty knots at full throttle.

FREQUENTLY ASKED QUESTIONS ABOUT LOBSTER FISHING

Lobster fishermen get asked a lot of questions. People "from away" are quite curious about the process of catching lobsters and what happens during a day out at sea. Some questions asked of fishermen are quite intelligent, others are rather silly and a few are completely inappropriate.

FIVE QUESTIONS YOU SHOULD NEVER ASK A LOBSTER FISHERMAN

Fishermen are notoriously secretive about the elements of their success. To divulge exactly where they're fishing and what they're catching is to reveal their hand and threaten their livelihood. If you wander down onto a dock and engage a fisherman in conversation, there are five questions you should never ask. They are as follows:

1. How many lobsters did you catch today?
2. How many traps do you fish?
3. How much money do you make?
4. Where is the best fishing?
5. What's the best haul you've had this season?

THE SILLIEST QUESTIONS DOWN EAST LOBSTER FISHERMEN HAVE EVER BEEN ASKED

Beyond these five no-go questions, fishermen get asked a lot of silly questions from people unfamiliar with a life spent working on the sea. When I surveyed Down East fishermen for this book, they shared with me some wonderful gems, which I have listed below:

- Why do you park all your boats in the same direction in the harbor?
- Are they fresh (in reference to the freshly caught lobster)?
- Do you bring in all your traps every night?
- Do you ever watch the lobsters going into the traps?
- Does it hurt if one bites you?
- Do you catch Alaskan King Crab?

ANSWERS TO OTHER FREQUENTLY ASKED QUESTIONS ABOUT LOBSTER FISHING

The contents of this book have, in many ways, been inspired by the questions I am frequently asked by people curious about lobster fishing. While I will cover the answers to these questions in subsequent chapters, below are shortcut answers to some of the more popular lobster-fishing FAQs.

- **How do you find your traps?** A fisherman uses uniquely colored buoys to mark the placement of each trap. Navigational equipment or landmarks are used to guide the fisherman to his buoys.
- **How many lobsters do you get in one trap?** The number of lobsters a fisherman catches in a trap will vary dramatically based on the time of year and how the lobsters are running in a certain area. In the spring especially, it's not uncommon for a trap to come up with no lobsters at all. In the good fall fishing, you can catch as many as ten to twelve lobsters in a single trap.
- **How often do you haul your traps?** A lobsterman will usually haul each of his traps at least once a week, often several times a week. In peak fishing, some fishermen haul their traps every day.
- **How many traps do you haul in a day?** The number of traps a fisherman hauls in a day varies depending on the weather, tides

and time of year. On a light day, a fisherman may haul fewer than fifty traps. On a busy day, he'll likely haul several hundred.

- **What do you do with your traps and boat in the winter?** Up to 70 percent of Down East fishermen take up all their traps in the winter months, when the fishing is at its slowest. During this time, the fishermen will perform needed repairs on the traps and repaint their buoys. Some fishermen haul their boats ashore for the winter as well. This is especially the case for fishermen with older or smaller boats.

- **Why aren't lobsters red when you catch them?** When people think of lobsters, they most often think of the bright red creatures that are served up on dinner plates. But lobsters only turn red once they are cooked. While alive, a lobster's shell is a mix of many pigments and looks greenish-brown in color. During the cooking process, however, the heat destroys all the pigments save the red.

- **Do lobsters feel pain when you kill them?** Lobsters do not feel pain when you cook them, as they have no brain and just a very simple nervous system. In order for an organism to perceive pain, it must have a more complex nervous system.[22]

- **Do lobsters cry when you put them in the pot?** Lobsters actually don't have vocal chords or any other means of vocalization so, no, they don't cry. According to Dr. Robert Bayer, a professor of animal and veterinary sciences at the University of Maine and director of the research organization the Lobster Institute, any noise you might hear while a lobster is cooking is likely air coming out of its stomach through its mouthparts.

- **How often do you eat lobster?** Some people assume that lobster fishermen eat lobster every day. Yet each lobster a fisherman brings home to cook is money he must deduct from his weekly wage. So how often do fishermen indulge in lobster? Of the fishermen in my survey, almost 60 percent eat lobster less than once a month. Only 10 percent of the fishermen I surveyed eat lobster every week.

THE AMAZING AMERICAN LOBSTER

While a modern-day delicacy, the American lobster is little more than a bug. Classified as an arthropod, it sits in the same family as insects and spiders. Fresh out of the water, it is greenish-brown in color and rather gruesome looking. Yet this sea bug is remarkably robust and has many sophisticated features.[23]

The first thing most people notice when admiring a lobster is its two strong claws. Each lobster contains a beefier crusher claw for pulverizing shells and a smaller pincher claw with serrated edges for tearing soft flesh. The lobster uses these claws for self-defense, as well as for gripping and destroying its food. As with humans, lobsters are either right- or left-handed. Some lobsters will have the crusher claw on their right sides, while others will have it on the left. Lobsters spar with one another by entering into a "claw lock," where they grip each other's crusher claw and engage in a shoving match. If both lobsters are right handed, they will reach crossways and grip each other as though they are shaking hands. If one lobster is right-handed and the other left-handed, they will hold hands on the same side to duke it out.[24]

As with insects, lobsters are invertebrates. They have no internal bones but, rather, a hard outer shell, which protects their body like a suit of armor. In order to grow, a lobster must shed its old shell and grow a new one. This process, called molting or shedding, is a remarkable event. The lobster splits in two along its back and extracts itself entirely from its old shell—down to the feelers and eyeballs. The remaining old shell is a perfect double of the newly shed lobster. The new shell is paper thin, like jelly to the touch, and will take up to two months to harden.[25]

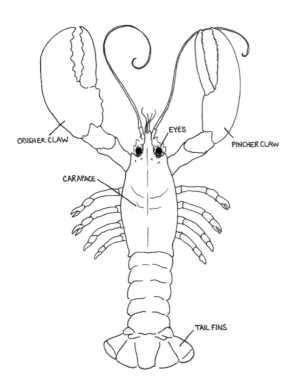

CRUSHER CLAW

EYES

PINCHER CLAW

CARAPACE

TAIL FINS

The American lobster. *Sketch by the author.*

In addition to molting in order to grow, a lobster is able to regenerate most of its appendages when needed. Lobsters often amputate their own claws and legs to escape danger. With time, a new claw will grow. Sometimes a deformed appendage will grow in place of another. For example, a leg may appear grotesquely in place of an eye. My father once caught a lobster that had two deformed claws growing out of its original claw joint. Apropos of appendages, the male lobster is equipped with two penises, with which it can complete "the deed" in a mere eight seconds.[26]

Though a creature of the sea, lobsters are capable of living outside of water for up to forty-eight hours in the right conditions. In the ocean, they can live to over one hundred years old. In fact, lobsters are such hardy creatures that scientists don't know the limits of their lifespan. As lobsters age, they don't slow down or become more prone to disease. They don't stop growing or become infertile. Lobsters actually become *more* fertile as they age. Because lobsters never stop growing, size is generally used to determine their age, though this is not a fail-safe approach, as lobsters grow at different rates depending on their environment.[27]

What we do know about the American lobster, more than anything else, is that it's a creature full of surprises. A simple bug that can outlive most humans. Bright red on your dinner plate yet greenish-brown in the wild. A poverty food of the yesteryears, now a symbol of fine dining. Let us hope this incredible icon of Maine continues to defy scientific predictions and thrive, for the sake of Down East fishermen and discerning diners the world over.

HOW TO CATCH A LOBSTER

Tricks and Tools of the Trade

TOOLS OF THE TRADE

lobster trap/pot: A hollow, rectangular cage approximately four feet in length designed to trap lobsters on the ocean floor. A length of rope connects the trap to a flotation device on the ocean surface, allowing the fisherman to find and haul the trap. Modern traps are constructed of plastic-coated wire mesh, while older versions were built of wooden laths.

bait bag: A mesh sack that holds the bait used to attract the lobster to the trap. Today, these sacks are made of nylon mesh, though originally they were knit from cotton twine. The bag is filled with the fisherman's bait of choice, most often herring, and strung up inside the kitchen of the trap, enticing lobsters to enter.

lobster buoy: A brightly colored, bullet-shaped buoyant marker that attaches to the trapline and helps the fisherman locate his traps. The streamlined shape of lobster buoys decreases the amount of drag against the strong ocean currents.

lobster boat: A motor-powered fishing vessel used to haul traps. Lobster boats can range from small, unpowered skiffs starting at fourteen feet to forty-eight-foot vessels with 1200-horsepower engines. Today, full-time commercial lobster fishermen are likely to have a diesel-powered, fiberglass boat somewhere between thirty-three and forty-five feet in length.

gaff: A long wooden stick with a metal hook on the end of it, which is used to gaff the line attached to the buoy so the fisherman can begin hauling in the trap.

pot hauler: A hydraulic winch with a rotating disk that pinches and pulls through the trapline, hauling the trap to the surface in the process.

snatch block: A pulley consisting of a simple roller system that the trapline runs through prior to being pulled through the pot hauler. This pulley keeps the line from rubbing on the side of the boat.

picking box: A container, often fiberglass, used to store lobsters that have just been picked out of the trap but still need to be measured and banded.

lobster gauge: A brass measuring device used by Maine lobstermen to determine if a lobster meets the minimum- and maximum-size regulations.

lobster bands: Thick, rubber elastics that are fitted over the lobster's claws to keep the lobster from pinching the lobster fisherman and other lobsters.

banding plyers: A metal implement, similar in appearance to regular pliers, that is used to slip the rubber bands onto the lobster's claws.

V-notch tool: An implement used to mark an egg-bearing lobster female for protection/conservation. The implement cuts a v-shaped notch into the tailfin of the lobster, identifying that lobster as unmarketable.

Tools of the trade: a lobster fisherman's banding box, filled with his banding pliers, bands and lobster gauge.

lobster tank: A container used to store lobsters and keep them fresh until the fisherman can return home and unload his catch. Today, most lobster tanks are made of fiberglass, plastic or aluminum, with circulating salt water flowing through them. Earlier lobster containers ranged from wooden boxes and metal washtubs to old wringer washing machines.

lobster crates: Plastic or wooden containers, the shape and size of a small cargo trunk, which fishermen use to store and transport their catch. As lobsters are stored and transported live, each crate has sections cut out to ensure proper ventilation.

lobster car: A large wooden floating dock with built-in compartments that store hundreds of lobsters, usually in crates. The lobster car is ventilated, allowing salt water to constantly circulate and supply the lobsters with fresh oxygen.

oil gear: The thick, orange outfits worn by fishermen to stay protected from the elements while working on the ocean. Most fishermen wear the Grundéns brand, which originates from Sweden and has a long-standing reputation for producing the best foul-weather gear. Until the beginning of the 1930s, these garments were dipped in barrels of boiled linseed oil

A lobster fisherman's uniform: thick Grundéns oil clothes.

to help make them water resistant. They are still referred to by fishermen as oil clothes or oil gear.[28]

trawl: A series of lobster traps connected to one another by a single line and fished as a set. Fishing a trawl allows a lobsterman to pull efficiently anywhere from eight to twenty traps to the surface with a single line.

knots: Lobster fishermen use a variety of knots in a typical day of fishing. Bowlines are often used for tow lines and attaching line to a float, while clove hitches are used for securing lobster boats to a wharf or piling. Water knots are used for connecting two sections of trapline or warp together.

THE ART OF THE LOBSTER TRAP

Modern-day lobster traps (also called pots) are rectangular, cage-like structures built out of coated wire mesh. Each trap is divided into two sections: the "kitchen" and the "bedroom." The kitchen end of the trap contains several funnel-type entrances or "heads" made of netting, through which the lobster enters the trap. The lobster is motivated to crawl into the trap because inside the kitchen lies a bag of enticing bait. After taking the bait, the lobster will seek to exit the trap, but the design of the trap tricks the lobster into crawling into the trap's bedroom. Essentially, the kitchen end of the trap contains a false exit. In addition to the inward-facing heads, there is another large and easily entered funnel (or head) that leads to the bedroom. The lobster will crawl up this head, thinking she's leaving the trap. Once the lobster passes through the very narrow end of this false exit, she finds herself trapped in the bedroom.

The bedroom does contain several small exit vents—called escape vents—that allow small, undersized lobsters and other sea life to exit the trap. In Down East Maine, traps are also required to have a biodegradable "ghost panel," which will degrade over time as it's left in the water. In the event the fisherman loses a trap, meaning the trap becomes separated from its buoy and line and is therefore stuck on the ocean floor, any sea life trapped in the bedroom will eventually be able to escape.

A wire trap is light in and of itself, so in order to ensure that the trap stays put on the ocean floor, the fisherman will fill the kitchen end of the trap with anywhere from twenty to fifty pounds of ballast (more in some cases)—typically cement, bricks or iron weights. The ballast also ensures that the trap lands the right way up when it hits the ocean floor. Once ballast has been added to a trap, it has considerable weight, and if it catches the fisherman on its way overboard, it can haul him straight to the bottom.

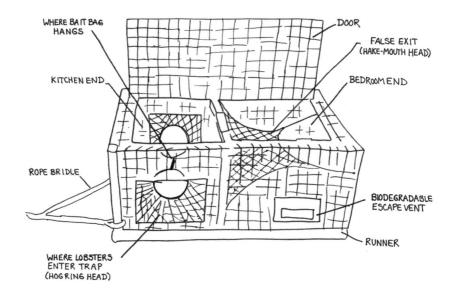

WHERE BAIT BAG
HANGS

DOOR

FALSE EXIT
(HAKE-MOUTH HEAD)

KITCHEN END

BEDROOM END

ROPE BRIDLE

BIODEGRADABLE
ESCAPE VENT

RUNNER

WHERE LOBSTERS
ENTER TRAP
(HOG RING HEAD)

The workings of a wire lobster trap. *Sketch by the author.*

When it's time to set the trap, the fisherman will measure a section of rope long enough to stretch from the trap, once planted on the ocean floor, to the surface of the water. Depending on where he plans to set the trap, this line may be anywhere from ten to one hundred fathoms in length. The fisherman will tie one end of the line to the trap and the other end to the buoy. Once the trap is in the water, the buoy will remain on the surface, marking the location of the trap and allowing the fisherman to haul the trap back up. The fisherman will mark the trap as his own by adding a government-issued tag with his unique fishing number.

When fishing in deeper waters of forty-five to sixty fathoms, some fishermen prefer to string a series of traps together into a trawl. Fishing a trawl is an efficient way of tending traps in deeper water, as you only have to find one buoy and haul one length of trapline to get to a number of traps.

Lobster traps vary in size, from thirty-six to fifty-four inches long. The trap size deployed by an individual fisherman will depend mostly on his personal preference. While logic would lead one to think that a bigger trap equals more lobsters, some fishermen find that a smaller trap fishes better. Other fishermen prefer a smaller trap as it's easier to lift and therefore less wear and tear on the body.

Today's wire lobster traps are relatively standard in shape and design, as they are mass produced by commercial trap builders. Prior to the mid-1980s, however, lobster traps were primarily made of wood and built by individual fishermen. Back then, the design of a fisherman's traps was based on his knowledge and skill, and the size of his trap gang was down to his dedication to trap building. A combination of ambition and talent could give one fisherman a big competitive advantage over another. A smart fisherman would tweak his design over the years, experimenting to see what alterations would lead to better fishing, while a fisherman with less skill would stick to the same design. An ambitious fisherman would work steadily through the winter months, building new traps for his gang, while a more idle fisherman would take his time and end up with a smaller gang of traps in the spring. Even the heads of traps were hand-knit by fishermen and their wives in the days of wooden traps. In fact, my mother went into labor with me one evening while she was knitting heads.

Once wire traps came on the market, professional trap builders began mass producing them. These trap builders would often work with a skilled fisherman to develop a design template that ensured good fishing. All of a sudden, a fisherman could write out a check and walk away with five hundred new traps in an afternoon. The fisherman didn't need to know anything about trap building. He didn't even need the actual money to purchase the traps. Often he'd just get a loan from the bank to buy his gang, with the intention of paying the bank back once he'd made his money later in the

My grandfather out fishing
during the days of wooden traps.

season. Between 1967 and 1997, the DMR estimates that the number of traps in Maine coastal waters more than doubled. Much of this increase can be attributed to the advent of wire traps. When asked what was the biggest change in the lobster industry since they started fishing, many of the older fishermen in my survey mentioned the vast increase in traps being fished.[29]

THE BOAT

An Overview

Lobster boats can range from small, unpowered skiffs starting at fourteen feet to forty-eight-foot vessels with 1200-horsepower engines. Today, full-time commercial lobster fishermen are likely to have diesel-powered, fiberglass boats somewhere between thirty-three and forty-five feet in length. Boats on the smaller end of the scale have the advantage of being able to maneuver easily in shallower waters, where lobsters love to hide. Smaller boats also cost less to build and operate. Larger boats benefit from the ability to carry bigger loads and handle rougher seas. When taking up and setting off gear, a larger boat can carry many more traps in a single load, which equals fewer trips. A bigger boat is almost essential for fishermen who fish most of their gear in offshore waters, where sea conditions can change quickly.

The design of a Down East Maine lobster boat is a balance between seaworthiness and functionality. The boats have a high bow, making them relatively seaworthy when heading into the wind. The stern and sides are wide and shallow so that lobster traps can easily be hauled aboard and wind action is minimized when the boat is broadside. Wind action is also minimized by the boat's long keel, which runs almost the full length of the hull below the waterline. The keel and hull work together to ensure that the boat tracks well in a following sea, one of the most difficult sea conditions in which to handle a boat. The hull itself comes in one of two designs: built-down or skeg. Both designs have unique advantages, and individual boat owners will attest to the superiority of their boats over others.

The boats have a semi-enclosed wheelhouse, which provides a modest amount of protection from the elements. This is where the fisherman and sternman stand throughout most of the day. The wheelhouse is kitted out with electronics and other equipment to help the lobster fisherman navigate, plot his gear and increase his safety. It also contains a hydraulic pot hauler, which the fisherman uses to pull up his traps from the ocean floor.

A harbor full of modern lobster boats.

In front of the wheelhouse is a rather small cabin or cuddy where the fisherman can get access to the engine and where other essentials, such as engine oil, life jackets and spare gloves, are stowed. Many boat designs include a rudimentary V-berth, but these beds are mainly used for storage rather than sleeping. As lobster fishing is mostly a day fishery, the boats are usually not kitted out with creature comforts such as stoves, freshwater tanks or toilets. For a man, the lack of a toilet is not so much of an issue. As a woman, I found it one of the hardest things about working on a lobster boat. When nature called, it meant hovering over a five-gallon bucket down forward, trying to keep my balance and my dignity as the boat bucked through the waves.

Beyond the basic design of a high bow, shallow stern, wheelhouse, cabin and unfortunate lack of toilet, each lobster boat has a unique configuration reflecting the needs and personality of the owner. Most boats are custom built to the captain's size and shape, so a boat built for a five-foot, seven-inch fisherman will look different than one built for a person measuring six feet, two inches. In some instances, the boat of a left-handed fisherman is rigged opposite to that of a boat for a right-handed fisherman. For a right-handed fisherman, the steering wheel, pot hauler and electronics will be rigged on the starboard side. A left-hander, who wants to gaff his buoys with his left hand, will rig his boat port side. Money also factors into the construction. Two fishermen can place an order on the same hull but end up with entirely different-looking boats if one has more cash to pour into the project. Most Maine lobster boats are built in Maine boatyards, though a few hail from Nova Scotia and are often referred to as Novie boats. Jonesport-Beals in particular has a long-standing reputation for building swift, seaworthy boats with graceful lines.

"How much does a lobster boat cost?" is a question often asked of lobster fishermen. While the cost of a boat will vary depending on the size and finish, most fishermen will say that their boat is their largest single investment. A forty-two-foot fiberglass boat with a decent motor in it today will cost over $200,000. This is often more than the price of the fisherman's house. The fisherman handles this colossal financial burden by taking a loan out from the local bank, usually for up to 80 percent of the boat's worth, putting his house and vehicle down as collateral.

A fully kitted fiberglass boat is a considerable investment for a man in his early to mid-twenties. It's also a fairly risky investment given that lobster fishing is, at best, an industry prone to instability. There are good years and there are bad years. Some years the lobsters are scarce. Other years, the lobsters are abundant but the market price is depressed. Even when the lobsters are flowing and the market price is good, the cost of bait and fuel can become disproportionately inflated, chipping away at the fisherman's margin. When times get tight, there are repossessions. A smart fisherman will work his way up slowly, starting his career in a smaller, less costly boat as he builds up his experience and savings. When he moves on to a more expensive boat, the experience and savings should allow him to ride out the rougher years.

A Boat by Any Other Name

One of my favorite things to do when gazing at a harbor full of lobster boats is to study the names. The *Vanessa Jane. Money Pit. Freedom.* Each name tells a story. A wife or daughter. A financial commitment. Liberation from corporate America.

Following a long maritime tradition, most lobster boats are named after women. In fact, roughly 60 percent of lobster boats in Down East Maine have a female name. Often the lobster fisherman will name his boat after his wife or daughter. This was the case in my family. My father named his first thirty-three-footer the *Celia Marie* after my mother. His subsequent boats have been named the *Christina Marie*. My survey of Down East fishermen showed that 32 percent had named their boats after their wives, 18 percent had named their boats after their daughters and 9 percent had named their boats after their granddaughters.[30]

Sometimes "Miss" will be used as part of the name to signify a daughter or granddaughter. The *Miss Emily*, *Miss Ginny*, *Miss Alayna* and *Miss Christy* are currently bobbing up and down in the harbors of Washington County. In jest, you can also find the *Miss Behavin'*, the *Miss Fortune* and the *Miss Guided* along the coast of Down East Maine. If a fisherman happens to have two or more daughters, it can get a bit tricky. That's when you end up with names like *Daddy's Girls* and *Four Daughters*. Other female boat names tell an even deeper story, such as *Becky's Worry* of Blue Hill and *Beth Said Yes* of Bar Harbor.

Beyond female names, there are several other naming themes for Down East Maine boats. Many boat names reference the significant financial undertaking of owning a lobster boat. *Under Pressure*, *All In* and *Pretty Penny* are reminders that today's lobster boats cost more than the average house in Maine. *Debt Ication*, *Second Mortgage* and *A Loan Again* hearken to the important yet precarious relationship most fishermen have with the bank.

Yet for every *Rock Bottom* there is a *Rich Returns*, for fishermen are also likely to pin their boat names to the promises of deep-sea riches. *High Hopes*, *Silver Lining* and *Rich Ambitions* bob up and down alongside *In the Red* and *Empty*

John Chipman Jr., a forth-generation lobster fisherman from Bunkers Harbor, and his boat, the *Theresa Anne*, named after his wife. *Courtesy of Theresa Chipman.*

My brother, Nick Lemieux, age thirteen, getting ready to launch his first boat, aptly named *Nick's Future*.

Pockets in Down East harbors. Younger fishermen will often name their boats after their industry calling. For example, my brother's first boat was named *Nick's Future*.

Other names stretch beyond the language of optimism to touch on the nonconformist character of Down Easters. *Born Free*, *Independence* and *Who's the Boss* are gentle nods to the rugged individualists who reside in this easternmost point of America. *Outsider*, *Outcast* and *Bad Company* are more extreme examples of the defiant nature of some Down Easters.

A final boat naming theme that bears mentioning is the perhaps predictable nautical puns, such as *Nauti Gal III*. The more lobster-focused range from *Chasin' Tail* and *Lobster Mobster* to *Entrapment* and *Standin n Bandin*.

A DAY OUT ON THE LOBSTER BOAT

Up Before Sunrise

One of the first things to know about a day out fishing is that it starts early. Most fishermen will be up and out the door no later than 5:00 a.m. Often it's much earlier. They start early to try to beat the wind, which tends to pick up in the afternoon. They also will likely be logging a ten- to twelve-hour day on the ocean, and an earlier start means an earlier finish. For fishermen working inshore, the tides will often play some role in when they head out to traps.

Checking the Conditions: Tide, Wind and Fog

Before leaving home, a fisherman will always check the tides and the weather conditions. The night before, he'll consult his tide calendar and watch the local weather forecast. In the morning, he'll check the direction and speed of the wind and the thickness of the fog, if there is any. These factors will affect how he approaches his day of fishing and whether it's viable to fish that day at all. He'll either check his weather vane, tune into the local weather station or call out to another fisherman who's already out on the water via VHF to ask, "How is it out there?" Because the conditions inshore often differ from the conditions on the water, consulting another fisherman is the one sure way to know whether it's worth heading out. As long as it's not blowing a gale or thick o' fog, he'll probably go.

A Day Out on the Lobster Boat

Lobstermen fish through all sorts of weather. They go out when it's raining. They go out when it's snowing. They go out when it's subzero. Their thick, waterproof Grundéns protect against extreme weather, and a special hat, called a sou'wester, can be worn to wick away extreme rain. At times, I actually enjoyed fishing in the rain when I was working on the boat. There was something about seeing the wet weather all around me yet being completely protected inside my Grundéns that almost made me feel cozy. The sight of raindrops falling onto the ocean and dissolving into the waves is also quite beautiful. Even sleet and snow often won't deter a fisherman from going to haul. High winds and thick fog are the most likely reasons a fisherman will skip a day on the water.

It's quite obvious why fishermen are reluctant to fish in high winds. Such conditions create high waves that make it challenging, and sometimes dangerous, for the fisherman to tend his gear. Rough seas make it perilous to work among the shoals and ledges where many fishermen place a portion of their traps. Shallow, rocky inlets are ideal spots for catching lobsters but tricky places from which to retrieve lobster traps unless the seas are calm. I remember several spots along the shore where my father would set a trap that we would only be able to haul every few weeks, when the weather was just right. Southerly winds are especially precarious for fishermen working along the Maine shores, as there is less room for error. If a fisherman's engine breaks down or he gets rope in his propeller during a southerly wind, it can quickly blow him onto the shore.

In the case of deeper-water fishing, high winds make it difficult to keep the boat stationary and stable while hauling. Seas are best handled by keeping the boat in motion and driving into the waves. When hauling traps, this is not always possible. You have to keep the boat in a fixed place to ensure the trap comes up well and that your trapline doesn't drift into that of another fisherman. If it happens that the wind is blowing crossways to the direction in which you have to haul, the boat can really take a kicking from the waves. While a fisherman's boat is almost always sturdy enough to take the thrashing, the fisherman himself is not. He can easily get caught out by a rogue wave, tossed against some equipment on the boat and end up with broken ribs or a dislocated shoulder.

At what wind speed will a fisherman hang up his hat for the day? It all depends on when the wind comes up. As the wind tends to pick up later in the day, fishermen are often already out hauling when things get blustery. In this case, fishermen will often put up with thirty-five to forty knots of wind in order to finish out their day. They'll just batten down the hatches and be extra careful while hauling. Before they head in, they'll call out to any remaining fishermen working the area to make sure those fishermen are okay. This is especially true if just a couple of fishermen remain offshore together in rough weather. Often they'll decide to head their boats for home at the same time so they can look out for each other along the way. When it gets to forty knots, things can happen fast, and you want someone close by if something goes wrong.

While a fisherman might finish out his day with forty knots, he would seldom venture out to traps in the morning with that much wind. Usually when the wind speed picks up to around twenty to twenty-five knots, a fisherman will question whether he'll make a day of it. In general, Down East fishermen are less likely to head out to haul in easterly and northerly winds, as they give the ocean more of a chop. A southwest wind tends to be an easier wind in which to fish.

I could write an entire chapter on Maine fog. It's the topic of many a VHF conversation and a subject about which fishermen love to moan. This is because we have a lot of it! Especially during the summer months, warm air reacts with the cold Down East waters, causing the moisture in the air to condense. Fog is most prevalent when the winds are blowing southwest, as the breezes push warm air from inland down onto the coast. Some mornings, you can wake up to the whole harbor and town being shut in with fog. When

Skiffs at rest in the harbor on a foggy day.

this happens, the fishermen need to make a judgment as to whether or not to go out. Thick fog makes it hard to see one's gear and dangerous to navigate the boat. Though fishermen paint their buoys bright colors to show up in fog and have special electronic equipment to help them navigate in unclear conditions, fishing on a foggy day is still less than ideal. It's very easy for a boat or rocky outcrop to quickly "come up on you" when you're out in the fog, so extra caution must always be taken.

When the harbor is shut in with fog, a fisherman will call out to any other fishermen already on the water and ask if it's worth going out, as sometimes the fog is thicker inland than out on the water. A typical response from a fisherman might be that it's so thick "you can cut it with a knife" or that it's "as thick as mud." Thick fog is also often described as "pea soup fog." Some fishermen have a landmark on shore that they use to judge if it's too thick to go out. When I was growing up, there was a hackmatack tree two-tenths of a mile down the road from our house. On foggy mornings, my father would look out the kitchen window, and if he couldn't see the hackmatack tree through the fog, he wouldn't go out. Some summers the Down East coast will be shut in with fog for weeks at a time. When this happens, fishermen just have to pick the best of the bad days and plow through it.

HEADING OUT TO THE MAIN BOAT: THE SKIFF

If the tides are right, the fog not too thick and the winds not too strong, the fisherman will make his way out to traps. He'll haul on his boots, hop in his pickup and head down to the harbor where his boat is moored. On a good spring, summer or fall day, the harbor area of a typical Down East fishing village will be lined with pickups and the harbor will be empty of boats. There are few parking lots and even fewer parking wardens in Down East Maine, so the fisherman will just park his pickup along the side of the road or down on a wharf, leaving the doors unlocked and the keys right in the ignition.

The fisherman's sternman will usually meet him at the wharf, and together they will make their way out to the lobster boat. Again, most lobster boats are moored in the middle of a harbor or an inlet where there is protection from extreme weather, as well as sufficient depth of water to ensure the boats won't ground out at low tide. In order to get to and from the lobster boat, the lobster fisherman will keep a smaller boat, called a skiff or tender, moored to a float at the end of the wharf where he fishes.

Heading to the boat for a day of fishing. *Courtesy of Laurie Cates.*

Today, more and more fishermen are using fiberglass boats with outboards as their tender. These outboards make the process of getting out to the lobster boat quite quick and effortless. Merely tug on the cord to start the outboard, untie the line and off you go. Of the sixty fishermen I interviewed for this book, over half now use outboards as tenders.

Back when I was growing up and fishing with my father, more fishermen in my town used rowing skiffs to get out to their boats. There was something quite serene about rowing out to the boat in the early morning, hearing the oars dip in and out of the water. The harbor would be quiet, save for the cry of seagulls and the distant rumble of another boat engine firing up for a day of fishing. In summer, the sun would often be rising over the island, creating a postcard perfect scene. In the fall, cold air would create a layer of steamy mist hovering just above the water's surface. Everything felt very peaceful and still in these moments, but I knew I had a long day of work ahead of me before I'd be sitting in that skiff again to head back home.

On the way out to the boat, my father would do the rowing, but sometimes, at the end of the day, I'd row us back to the wharf. Only then could I appreciate the art of rowing and how hard it is to achieve perfect alignment and rhythm with the oars. I've spent years watching my father and grandfather row with great skill, and yet, to this day when I get behind a set of oars my stroke is always lopsided.

Often, older skiffs that have been retired as tenders make their way up to a fisherman's front lawn and become a playground for his children. In these older skiffs, kids will spend countless hours pretending to haul traps. This was the case with our family. My father's older wooden skiff was placed on our front lawn, and my brother and I used it to play "lobster fishing" when we were growing up.

Other times, old rowing skiffs are left down at the float to be used for a spot of exercise. Rowing can be a very pleasurable way of getting a workout, as it takes physical effort and coordination. If you ever get a chance to borrow a fisherman's skiff for a spot of rowing, just remember to put it back properly. Stow the oars under the seats to keep them secure, hinge down the oar locks so they don't dig against other boats and tie up the skiff with the same knot used by the owner.

Prepping the Boat and Plotting the Course

Once the fisherman gets out to his lobster boat, he'll fire up the engine. The process of starting up a lobster boat is similar to starting a car, performed by turning a key in the ignition. As with their pickup trucks, most fishermen never take the key out of the ignition. Once the engine is running, the fisherman will turn on his electronics and pump the bilge. He'll then unhook his boat from the mooring and head out to traps.

As he heads out of the harbor or inlet in which his boat is moored, the lobster fisherman will consult his radar, loran and Global Positioning System (GPS) satellite plotter equipment, along with his compass. Though modern electronics have made course-plotting easier, the humble compass is still a fundamental navigational aid. While working on the boat, I never learned to use the GPS or loran, but my father made sure I could steer the boat by compass. Before the advent of modern electronics, the compass was used, along with a clock and a fixed engine speed, to navigate in adverse conditions through the process of dead reckoning.

Along with the compass on his boat, the fisherman will usually pass several other classic navigational aids on his way out to sea, such as a channel marker or a lighthouse. These fixtures have become such icons of Maine, filling our postcards and picture books, that it's easy to forget the vital role they've played in helping bring fishermen safely home through the years. Down

East Maine alone has twenty-six lighthouses dotting its coast. One of these lighthouses is perched on an island at the head of my harbor. Called Little River Light, it was manned by my great-grandfather, Willie Corbett, during the early 1900s and by my great-great-grandfather, Roscoe Johnson, in the 1890s. Now automated, it continues to emit beacons of light, which help guide fishermen into our harbor in dark, stormy or foggy weather. Every lighthouse along the Maine coast is a unique height and has an individual flash pattern, called its characteristic. For example, the Little River Light tower is forty-one feet high and flashes every six seconds. Each fisherman is required to stow a lighthouse book on his boat so he can identify which light signal is associated with his home port.

As the fishermen steam past these icons, they'll be busily prepping the boat and themselves for the day. Bait will be brought out from under the stern or washboards of the boat, where it was stored to keep the gulls from getting at it. Grundéns oilskins will be removed from their hooks and put

Sunrise over Little River Light. *Courtesy of Bill Kitchen.*

on. Baskets will be put in place and tanks filled with circulating salt water to await the arrival of the lobster catch. Once the boat engine has warmed up a bit, the fisherman will open her up to cruising speed and steady his course for the first of his traps. The day of fishing is about to begin.

FINDING ONE'S TRAPS: THE LOBSTER BUOY

A common question asked of lobster fishermen is, "How do you find your traps?" While navigational aids and landmarks help to a degree, the lobster buoy is the key marker for a lobsterman's traps. Each fisherman has a distinct buoy color and pattern to distinguish his traps from those of his fellow fishermen. Buoy colors range from traditional reds, blues and greens to bright pinks and electric yellows. The brighter the buoy, the better chance it has of being visible through the fog. Buoy patterns often consist of lengthwise or crosswise stripes.

A shoreline littered with a kaleidoscope of different-colored buoys is one of the most beautiful and iconic images of Maine, yet this coloring serves a very functional purpose. Look closely at a harbor full of boats and you'll see that each lobster boat, and fisherman, takes ownership of a buoy color

My brother's lobster buoy color pattern. The color pattern was originally used by my grandfather. *Courtesy of Chad Hangen.*

pattern by placing a buoy with that said color pattern on top of his boat. Sometimes, this buoy color pattern will be handed down from one generation to another within a fishing family. For example, my grandfather's buoy color was orange with a blue stripe. When my grandfather finished fishing, my brother took over that color pattern and still uses it today. Today's lobster buoys are made of durable styrofoam and are bullet-shaped, while earlier versions were made of wood and shaped as spheres or rounded oblongs. You can find versions of these old-school lobster buoys in antique shops throughout Maine.

Hauling the Trap

Once the fisherman has spotted his buoy, he pulls up alongside it, cuts the engine and grabs ahold of the line directly below the bouy with a long wooden gaff. He then runs the line over the snatch block and down around a pot hauler. The pot hauler, a hydraulic winch with a rapidly rotating disk, pinches and pulls the line through, hauling the trap to the surface in the process. It's an efficient but also a very dangerous piece of machinery, as it can easily pinch your finger as well. Many fishermen have broken hands and

The snatch block and pot hauler, ready to pull up a trap. *Courtesy of Anthony Oragano.*

lost fingers due to catching their hands between the line and the pot hauler. One fisherman in Washington County watched the tips of his fingers drop to the floor of the boat after getting them caught in the pot hauler.

The fisherman must also be careful of his hands as he clears any tangles he finds in the line during the hauling of the trap. Tides, storms and sometimes sloppy fishing can lead the lines of several fishermen's gear to become tangled or snarled. When this happens, the fisherman has to untangle the two lines, which are extremely taut from the weight of traps and the pull of the tides. It's very easy for a fisherman to catch his hand or fingers in some of the tangled line. Several years ago, a Cutler fisherman had his thumb ripped off in one such incident.

Once the line has zipped through the spinning wheel of the pot hauler, it drops down to the floor and builds into a large coil, spewing slime and seawater around the cockpit in the process. Again, the fisherman keeps track of this coil of line to ensure that it stays neat and tangle free to avoid complications when the trap is eventually reset. At the same time, he's watching for the trap to surface, at which point he'll stop the pot hauler and haul the trap aboard.

Watching for a trap to surface is one of the most exciting parts of lobster fishing. The ocean in Maine is dark and murky, making it hard to see much more than several feet below the surface of the water. When I was working alongside my father, I'd often peer over the side of the boat, straining my eyes for the first glimpse of the trap rushing up through the water. Once spotting the first faint outline of the trap, the next thing I'd look for was a sign of lobsters in the trap. It was not until the whole trap was out of the water that I could clearly see exactly how many lobsters we had caught.

Landing a trap.

Picking Out the Trap

Once the trap is hauled up onto the gunwale, the trapdoor is opened and the picking begins. Often, the lobster fisherman picks the lobsters out of the trap while his sternman changes the bait bag. As he picks out the lobsters, the fisherman will toss overboard any that clearly do not make the legal requirements. Shorts, eggers and V-notchers are returned directly to the sea. All other lobsters are placed in a basket or picking box for the sternman to measure and band. The fisherman and sternman will also clear the trap of any crabs and fish that may have collected during the set. Unfortunately, you no longer find many edible fish, such as cod, in your traps, as these fish stocks have been depleted over the years. During the 1980s, I remember catching at least several cod a day, which my father and I would kill, clean and bring home to my grandmother for cooking or drying (dried codfish is an acquired taste but a Down East Maine delicacy). Currently, the inedible sculpin is the fish you'll most likely find alongside lobsters in your trap. This fish is usually killed and placed back into the trap as additional bait for the next set.

Picking out a trap. *Courtesy of Gillian Ryan.*

A Day Out on the Lobster Boat

Getting ready to set a trap. *Courtesy of Gillian Ryan.*

One of the more common questions you get about lobstering is, "How many lobsters do you get in a trap?" It's a good question but a hard one to answer, as the number can vary dramatically based on the time of year and how the lobsters are running in a certain area. In the spring especially, it's not uncommon for a trap to come up with no lobsters at all. In the good fall fishing, you can catch as many as ten to twelve lobsters in a single trap. A fisherman will usually only count the keepers, so if a fisherman tells you he's caught ten lobsters in one trap, it's likely that the trap contained four or five shorts and a V-notch along with ten lobsters meeting the legal size requirements. Conversely, a fisherman may say he didn't catch anything in a trap. In reality, that trap may have had several shorts or an egger, just no legal-sized lobsters.

Beyond catching a large number of legal lobsters, the most exciting thing to catch in a trap is a unique lobster. Blue and yellow lobsters are rare but have been caught by different fishermen along the coast of Down East Maine. Catching a big bull lobster is also quite exciting. The largest lobster my father has caught weighed around twenty-five pounds. The lobster was so large it actually could not fit through the heads of the trap. Rather, it had been straddling the trap and shoving its claws through the heads to get

Joshua Cates holding a supersize lobster he caught while hauling traps. The lobster was returned to the sea after being photographed and weighed (it tipped the scales at eighteen pounds). *Courtesy of Laurie Cates.*

at the bait when my father hauled up the trap. Because the lobster's claws had become stuck in the trap heads, the lobster rode with the trap to the surface of the water, giving my father a big surprise. Beyond big and brightly colored lobsters, mutant lobsters can also be found in a fisherman's trap. Though lobsters can regenerate most of their appendages, the regrowths don't always go according to plan. My father once caught a lobster that, in addition to its two normal claws, was growing two extra malformed claws out of one of its sockets.

While the fisherman is dealing with the often normal and sometimes unique lobsters, his sternman will take out the old bait bag and replace it with a bag of fresh bait. He'll empty the bait from the old bag over the side of the boat, where it's quickly scooped up by the circling gulls. He'll then close the trap and either get it ready to toss back over the side or, if the trap is part of a trawl, place it on the trawl table or out back to be run off the stern of the boat.

A Day Out on the Lobster Boat

THE SETTING OF THE TRAP

Once the trap, or string of traps, is all tended, it's ready for another set. Unless the fishing is very poor in that area, the fisherman will aim to reset the trap in the same area. If the trap is a single, the lobsterman will steam back to the spot where the trap was originally hauled and just push it off the side of the boat. The ballast of the trap creates a gravitational pull that sinks the trap to the bottom, pulling the line behind. The buoy is either tossed over or placed on the gunwale so it can easily slide off when the rope starts to pull. With a trawl, once all the traps are hauled and placed on the trawl table or the stern of the boat in the proper sequence, the fisherman will reestablish his coordinates and then kick off the trawl anchor as the boat is steaming forward. The force of the sinking trawl anchor pulling in one direction and the boat's forward motion in the other direction helps pull the traps off the trawl table or stern deck, one by one.

The process of setting a trap or trawl is one of the most dangerous elements of lobster fishing, and everyone on the boat must be vigilant during this time. When rope runs off a moving lobster boat, it flails across the deck before whizzing off the stern. In the process, it can quickly form a loop over anything in its path, including a hand or foot. I was never allowed to venture beyond the cockpit when trapline was running out. Some fishermen even keep a knife strapped to their oilskins in the event they get caught in the line as the trap is being set.

Another common question about lobster fishing is, "How long do you let a trap set?" Again, this depends somewhat on the time of the year and how well the trap is fishing. In most cases, a fisherman will leave a trap to set for less than a week. If he hauls his traps too frequently, he won't catch a decent number of lobsters. If he leaves the traps for too long, the bait will have lost its potency or been consumed by sea creatures, leaving no incentive for additional lobsters to visit the trap. So each week, the fisherman will typically make the rounds, hauling each trap of his gang once or twice.

In some instances, however, it makes sense to haul the traps more or less frequently. For example, at the very beginning and end of the season, when the fishing is really slow, it's often not worth it to haul every week. A fisherman may leave his traps to set for two weeks and they still come up dry. In spells of slow fishing, a fisherman may say he's going out to "change the water in the traps."

Conversely, in the fall, when the fishing really starts picking up, some fishermen begin "hauling back." Hauling back refers to hauling the same traps or trawls repeatedly in the same week. It's often worth the investment (fuel, bait and sternman wages) for a fisherman to frequently rehaul some of his traps when the fishing gets good. Once a trap gets a certain number of lobsters in it, it's less likely to attract more lobsters regardless of how fresh the bait is. Also, once a bunch of lobsters have been hanging around in a trap for a while, they often start fighting with one another, resulting in lost claws and dead lobsters. This is especially true with large egg-bearing lobsters, which are extremely defensive and will often chew up three or four smaller lobsters in the trap if they're left for a length of time. Hauling back helps ensure the lobsters are cleared from the trap before they get a chance to tear one another up and that the trap is primed to attract more lobsters.

SORTING THROUGH THE CATCH

Once the trap is set and the fisherman is steaming toward the next trap, he or his sternman will begin sorting through the catch to make sure all the lobsters are keepers. A seasoned fisherman will be able, in most instances, to tell if a lobster is a keeper just by eyeing it. To be safe however, any lobster that looks of questionable size will be measured. To measure a

Measuring a lobster. *Courtesy of Anthony Oragano.*

lobster, the lobsterman will take his brass lobster gauge and place one end of the gauge right below the lobster's eye socket. He'll then measure from the eye socket down to the back of the carapace, where the tail joins the body. If the measure extends beyond the carapace, the lobster is too small and must be returned to the ocean. Currently, in Maine, the carapace must measure at least three and a quarter inches in length for the lobster to be of legal size.

In addition to a minimum size for lobsters, a maximum-size law also exists in Maine. Any lobster with a carapace longer than five inches must also be returned to the sea. This maximum-size law, which currently exists only in Maine, protects the large, "breeder" lobsters and is likely one of the reasons the Maine fishing industry is so healthy. The rationale is that larger lobsters are capable of producing greater and healthier numbers of offspring. By returning all of the largest lobsters back to the sea, the fishermen are protecting this brood stock.

As the lobsters are measured, any that are short or oversized will be tossed over the side. The keepers will be double-checked for eggs and V-notches. Again, any lobster bearing eggs or with a notch in her middle right-side flipper must be returned to the sea as well. Any female carrying eggs will be V-notched before being returned to the ocean.

If the lobster passes the size and other conservation requirements, it's then banded. The main reason for banding lobsters is to ensure they don't attack one another while in the holding tank or lobster crates. The bands also ensure that the lobster does not bite the lobster fisherman or anyone else who may handle it. At one point or another, every lobster fisherman has been bitten by a lobster. And when you get bitten, it hurts! Lobsters have incredible strength in their claws—enough strength to draw blood or break your fingernail. I'm lucky to have only been bitten once by a lobster. I was holding the lobster with one hand and not paying enough attention to the other hand. Before I knew it, the lobster had latched hold of the skin right below my thumb. I was wearing gloves, but they did little to protect me. I still remember it as a very painful experience.

To band a lobster, the fisherman will clutch the lobster at the base of its claws with his nondominant hand and, with his other hand, grab the lobster pliers, scoop up a band and slide the band down over the lobster's claw. Once the band is about three-fourths of the way down the lobster's claw, the fisherman will release his grip on the pliers while twisting his wrist in a quick, steady motion. This process releases the band from the pliers and plants it firmly over the lobster's claw.

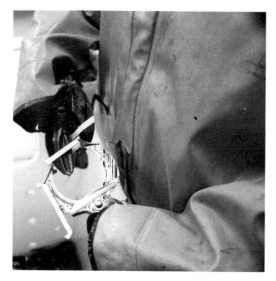

Banding a lobster. *Courtesy of Anthony Oragano.*

Banding lobsters looks easy until you try to do it yourself. A seasoned fisherman has banded so many lobsters in his day that he'll do it with incredible speed and finesse. When a newcomer takes hold of the banding pliers, however, he looks similar to a young child learning to tie his shoes. Bands will be dropped or fired across the boat as he tries to operate the pliers. Once he gets the band down onto the claw, it's quite a struggle to pull the pliers away without the band coming back off.

Lobster bands (and banding pliers) have been used since about the 1950s. Prior to this time, lobster "plugs" or "pegs" were used to keep the claws in check. Lobster plugs were narrow wedges of pinewood that the lobsterman would jam into the thumb joint of the lobster's claw, preventing the claw from opening. As lobster became increasingly popular, however, bands replaced plugs as a more humane and hygienic option (once a lobster's shell is punctured, it can quickly become diseased).

Once banded, lobsters are placed in a ventilated basket or holding tank. In either case, circulating salt water is used to keep the lobsters fresh until the end of the day of fishing. When I was fishing with my father, I spent many hours staring down into his saltwater lobster tank, checking to see how many lobsters were collecting throughout the day. If we were able to cover the bottom of the tank with lobsters by 10:00 a.m., it usually meant it was going to be a good day of fishing. Other days, covering even the very bottom of the tank with lobsters seemed an impossible task. As my pay was directly linked to the volume of the catch, a day of slow fishing was especially depressing.

In addition to sorting the catch between traps, the fisherman or sternman must also prep another bait bag. Baiting up is one of the nastiest tasks in the world of lobster fishing. Stuffing foul-smelling herring into a mesh sack, especially on a foggy day with a rolling sea, is not for the weak of stomach. While I've always had a good pair of sea legs, the constant sight and smell of lobster bait in the wrong conditions could make me sick as a dog during a day of fishing.

Finishing Up the Day: Scrubbing the Decks and Unloading the Catch

On a busy day, fishermen can be on the water for twelve hours or more. Most of this time is spent hauling traps, with the last hour and a bit devoted to cleaning the boat and crating and selling the catch. It's a very exhausting day, as you're on your feet the whole time, doing lots of lifting and pulling out in the fresh sea air. Lunch is often consumed standing up while steaming between traps. The process of pulling trap after trap can also feel quite monotonous.

Fishermen are often asked, "How many traps do you haul in a day?" It varies, depending on the weather, tides and time of year. On a light day, a fisherman may haul fewer than fifty traps. On a busy day, he'll likely haul several hundred. Another very common question is, "How many lobsters does a fisherman catch in a day?" The answer to this is, "Don't ask." One of the least acceptable subjects of conversation in the world of lobster fishing is how much a fisherman is catching. If you ask a fisherman how much he caught after a day on the water, he'll likely answer, "Oh, a little more than yesterday," without explaining exactly how much he caught yesterday. It's a polite way of skirting the issue.

The only time a fisherman will talk about how many lobsters he's catching is if the volume is so low it's noteworthy. For example, a fisherman might get on the VHF and complain to his fellow fishermen that he's hauled ten traps for just one lobster. During spells of poor fishing, the VHF will be alight with fishermen lamenting the lack of lobsters. Sound bites might include "I've fished all day just to break even," "I'm out changing the water in the traps" or "I could do better fishing in a mud puddle." If you tune in the VHF during a streak of good fishing, you'll probably need to recheck that it's actually on. All the fishermen will be busy tending to the gear and the influx of lobsters, and any conversation will stick to the essentials—tangles, missing gear and estimated homecoming times. In Down East Maine, there is an inverse relationship between the amount of VHF conversation and the number of lobsters being caught.

Once the hauling is done and an undisclosed volume of lobster is resting in the holding tank, it's time to head for home and start scrubbing down the boat. Liquid dish detergent mixed with water in a bucket is as good a cleaning aid as any to scrub away the greasy bait drippings, seaweed and other mess that has collected throughout the day. While the lobsterman is steering the boat for the harbor, the sternman will take a pail of diluted detergent and start scrubbing the decks. He will also tidy the boat by gathering stray empty bait bags, coiling up loose rope and stacking and stowing empty baskets and crates. Often, this work will take as long as the trip home. When I was fishing with my father, I relished those rare times when I was able to finish scrubbing the decks before we got to the harbor. I would take those precious moments to sit on the stern deck, close my eyes and feel the wind and ocean spray on my face. I could always tell, without opening my eyes, when we'd reached the harbor because the air would immediately become warmer and full of the sweet smell of evergreen trees.

Crating up the catch. *Courtesy of Gillian Ryan.*

A Day Out on the Lobster Boat

Above: Weighing up a crate of lobsters. *Courtesy of Laurie Cates.*

Right: One of our village's fuel pumps, where lobstermen fuel up for a week of fishing. *Courtesy of Chad Hangen.*

When the fisherman gets into the harbor at the end of the day, he'll head to the wharf from which he fishes and start unloading his catch. The wharf owner or a wharf hand will usually greet the fisherman and help with the process. The lobsters will be transferred from the holding tank or baskets into ventilated crates in units of ninety to one hundred pounds. As lobsters are sold based on weight, having a consistent weight for all the crates makes the selling process easier and more organized.

Traditionally, the crates were made of wooden slats, with spaces between the slats to allow for the circulation of water and air. These wooden crates posed a challenge, however, in that they had no fixed weight. The wharf dealer would have to weigh each crate individually and subtract the weight of the crate before loading it up with lobsters. Some scales found on wharves today still have a double set of balance beams used for balancing the weight of the crate against the weight of the lobsters.

At some point in the 1990s, wooden crates were replaced by more efficient plastic crates. You can still see the older wooden crates during Fourth of July or Harbor Day celebrations along the coast as part of the unique Maine sport of "crate racing." In addition to wooden crates, metal

My grandfather, Glenn Farris, handling lobsters on his wharf. In his right-hand shirt pocket you can see the thick wad of cash he carried and used to pay fishermen on the spot for their catch.

washtubs or scale baskets were often used as a means to pass lobsters from fishing boat to wharf.

Now, if you wander down onto a wharf at the end of a day of lobstering, you'll see fishermen and wharf hands filling modern, multicolored plastic crates to the top with lobsters and then placing the crates on a scale and adding or removing lobsters to get to an exact weight. The wharf owner will keep a tab of the total poundage from the day. He'll usually wait until the end of the week to cut the fisherman a check for the week's catch, though there are instances where a check is cut on the day. Back in the mid-1900s, when my grandfather was a lobster dealer, fishermen were paid in cash, on the spot, for their catch. In his front pocket, my grandfather would always carry a big wad of cash to give his fishermen in exchange for the lobster they caught that day.

After unloading the catch, the fisherman will take on fuel, bait and any other supplies needed for his next day at sea, moor his boat and head home for a well-earned supper.

A LOBSTER FISHERMAN'S JOB IS NEVER DONE

As a self-employed person, a lobster fisherman's job is never done. Though the traps may be hauled for the week and the boat decks scrubbed, the fisherman will often be occupied with boat maintenance, bait deliveries and more before he is able to take off his boots and put up his feet.

BOAT MAINTENANCE

Anything that sits for a while in the nutrient-rich waters of Maine quickly becomes covered by a film of dark green algae. This is especially true of the underside of lobster boats. About every two months, a fisherman will clean the bottom of his boat to ensure the sea growth doesn't impact the boat's performance. The intense tides of Down East Maine help the process by creating two opportunities each day for a fisherman to "ground out" his boat and get the work done.

The process of grounding out a boat starts at the beginning of ebb tide and involves either tying the boat to the side of a wharf with bow and stern lines or driving the boat up onto a sandy beach until the keel is touching the sand. Ballast is often used to ensure that the boat lists in the proper direction. The fisherman then just waits for the tide to go out. Within two to three hours, the boat's undersides will be clear of water and he can access the hull with ease. Today, high-pressure washers are often used to spray away marine growth. In the past, bleach was used. While the boat is grounded out, the

Cleaning the bottom of the boat at low tide.

Getting rope out of the boat's propeller.

fisherman will also check the propeller, removing any rope that may have collected in the wheel.

While the above description makes the process of grounding out a boat sound simple and matter of fact, there have been many instances in which boats have toppled over onto the beach or rocks because the fisherman forgot to ballast the boat or didn't understand how to tie it off correctly. Additionally, a boat lying in the sand fifteen feet up a beach can look rather odd to a person not familiar with our tidal system. One visitor to Cutler who saw my father's boat ground out on the beach exclaimed with awe, "You must have had her going wide open to drive her way up there!"

TAKING CARE OF BAIT

Anyone who has walked down onto a working lobster wharf in the summer months will be familiar with the potent smell of lobster bait. Made up primarily of herring, this bait requires a fair bit of handling and care before it makes its way into lobster traps.

Unloading lobster bait from the bait truck.

A Lobster Fisherman's Job Is Never Done

Salting the bait to preserve it while it's stored on the wharf.

Transferring lobster bait to totes for a day of fishing. *Courtesy of Mike O'Connell.*

Today in Down East Maine, the bait is mostly delivered to fishing wharves in big eighteen-wheelers. The bait trailer is equipped with a chute from which the foul-smelling herring spills out, filling fifteen-bushel plastic Exacto boxes. As the bait is transferred from trailer to the insulated Exacto boxes, the fishermen coat it with salt to preserve it. Once each box is full, it is sealed and stored on the wharf until needed for fishing.

Some fishermen employ a person to fill their bait bags (bait up) while the bait is still on the wharf. One man in our town who baited up bags for a number of the fishermen was jokingly crowned the town "master baiter."

Once per week, usually on a Sunday afternoon before the next week's lobster fishing begins, the fishermen will transfer their bait once more from the large Exacto boxes into smaller totes and lower the bait down onto the lobster boat. On the boat, it's stored under the stern or decks, out of the reach of the gulls but handy for the fishermen to use as needed.

THE STRATEGY OF
LOBSTER FISHING

To an outsider, lobster fishing could appear to be a simple, possibly dull exercise of navigating a boat and methodically hauling trap after trap. For a fisherman, maneuvering the vessel and pulling traps are mechanical tasks, performed almost automatically while he focuses on the more important job of deciding where to place traps to maximize his catch and minimize trap tangles or damage. His thoughts are on the possible movements of the lobsters and the movements of other fishermen. As he places his traps, he will consider a vast range of factors, including how lobsters behave in different seasons, the weather conditions, the activity of the tides and the cost of his expenses (such as the fuel he will use to get to the traps). He will note the trap placement and hauling patterns of fellow fishermen not only to minimize entanglements but also to try to gain a competitive advantage. Over the years, he will tinker with the design of his traps. Throughout the season, his head will constantly recalculate potential profit margins based on the ever-changing cost of his supplies versus the volume and value of his harvest.[31]

While lobster fishermen spend countless hours on the VHF or gathered around a pickup truck, talking about the weather, government regulations and gossip, they spend many more hours quietly pondering how to optimize their fishing performance. Throughout the season, a fisherman will toy with his trap placement and frequency of hauling to maximize his yield from the ocean floor. He may also experiment with his trap design and the bait he uses. Though hard work and dedication go a long way, knowledge and skills are critical for bringing in a decent catch. Most harbors have at least one fisherman who is renowned for his ability to catch lobsters, regardless of the size of his boat or the level of his ambition. In fact, knowledge and strategy are so important that fishermen are incredibly secretive about their tactics. To reveal where they fish and how much they catch would be to show their hand and threaten their livelihood.[32]

With regard to trap placement, a fisherman will aim to place his traps where he judges they will catch a maximum number of lobsters with minimal chance of damage or loss. A skilled fisherman will not place his traps too close to the shore, where they can be swept up onto the beach or ledges during foul weather. At the same time, he'll be adept at working around the ledges where lobsters love to hide. When fishing farther offshore, a good fisherman is proficient at knowing the exact measure of warp needed to keep his buoys watching well in the tides while not tangling with neighboring traps.

Most importantly, a skilled fisherman knows how to stay ahead of the lobsters and, at times, the other fishermen. Just after the lobsters shed, a skilled fisherman will place most of his traps in bays and other shallow areas inshore or on shallow ledges offshore. When lobsters begin to migrate offshore in the late summer, a smart fisherman will move his traps to gravel bottom or along the edge of hard bottom, keeping pace with where the lobsters are "running."[33]

In order to stay ahead of the lobsters, a skilled fisherman will also place a percentage of his traps in areas where he does not expect to immediately catch lobsters, simply to reserve the spot or to scout out the ground. While the current fishing might be most fruitful in the shoals, by fishing a portion of his traps on gravel and hard bottom, he'll be the first to know when the lobsters start running in that area. If he's able to catch the beginning of that run without the other fishermen discovering it, he'll catch a higher yield of lobsters.

Beyond the placement of the traps, the design of the traps themselves and the leveraging of bait are probably the next most important factors that a savvy fisherman will employ to maximize his catch. While there is a fair amount of consistency in the general design of today's traps, small nuances can make a difference. For example, different fishermen leverage different styles of heads to lure lobsters into the kitchen and the parlor of the trap. On the traps of most experienced fishermen, you'll see round hog ring heads for the kitchen entrance and then a narrow hake-mouth head for the entrance to the parlor.

A smart fisherman will also play close attention to the working time of his bait. The quality of the bait itself, the temperature of the water, the volume of lobster moving across the bottom and the activity levels of other sea life on the ocean floor have an impact on how long a bag of bait will effectively attract lobsters to a trap. Certain conditions can lead bait to lose its potency or, as fishermen say, "go off" more quickly. In Down East Maine, easterly winds tend to stir up sea fleas, which eat away at the bait, leading to empty bait bags and less effective fishing. When the bait is not staying on well, a shrewd fisherman will be good at judging whether the expense of adding an extra haul to refresh the bait will be offset by the additional lobster he'll harvest in the process.

Hauling more frequently does not always equal more lobsters. A skilled fisherman will understand the impact water temperature plays on lobsters' behavior and plan his fishing accordingly. Often in colder water, lobsters move more slowly, and an extra haul can be counterproductive. Understanding such nuances helps a fisherman maximize his yield with minimal expense.

Accomplished lobster fishermen don't rely completely on their electronic equipment. They commit good lobstering areas to memory, working the same "honey holes" year after year. They mark their inshore positions by landmarks versus GPS. Some of these landmarks can have quite colorful names. For example, along the shores of Cutler's fishing grounds you'll find the landmarks of Wet Ass, Shittin' Brook and Pig Turd Point. How they got these names is a story for another time!

Finally, smart lobster fishermen don't rest on their laurels. No season is the same. It's a game of constant experimentation. When looking through the MIT research on Down East lobster fishing, I came upon the following quote, which sums it up perfectly:

> *Fred was sitting on his doorstep when a guy with New York plates come up and says he wanted to know all there was to know about fishing. What Fred told him was that all he knew was that it changes from year to year—you can fish one area one year and do real good—and next year there may be nothing…All I can tell you is that from year to year things change, but we somehow make a living—my philosophy is to spread my traps around—not put all your eggs in one basket. No one man can tell you all there is to know about fishing.*[34]

WORKING THE TIDES OF DOWN EAST MAINE

While in most senses the strategy of lobster fishing in Down East Maine is similar to that of fishing in the rest of Maine, one chief difference is the force of tide with which Down Easters must deal. Tides in the eastern tip of Maine reach upward of twenty-seven feet, much bigger than the typically nine-foot tides of Maine's southern coast. The highest tides in the whole of the United States can be found at the very tip of Down East Maine, where the water rises or falls three-quarters of an inch with each passing minute. The velocity of the tides is so dramatic that a number of people have drowned after being caught out in the quickly rising tides. The powerful tides of Down East Maine also have profound implications on how lobstermen approach their fishing, for the tide cycles impact not only how their traps fish but also whether those traps can be fished at all.[35]

Four times a day, the ocean waters of Down East Maine rush northeast, filling the Bay of Fundy, or southwest, emptying out the harbors and coves along the coast. In the process, they pull with them the warp lines that link traps on the ocean floor to buoys on the ocean's surface. The sheer force of the tides is so great that even the brightly colored buoys that fishermen use to find their traps are pulled under water in the process. Even though fishermen in some parts of Down East Maine fish up to three times the length of line needed to reach from their trap to their buoy, the force of the flood and ebb tides is still great enough to drag the buoy under. If the buoys aren't showing, it's impossible for a fisherman to haul his gear. Therefore, Down East fishermen must work around the tides, conducting their fishing at the slack points between high and low water, when the tide is relaxed and the fishermen's buoys are showing on the surface of the water. Lobstermen in Down East Maine will speak of "going out for a slack" or "catching a slack." These fishermen are referring to timing their day of fishing so that they arrive at their fishing ground when the tides are still and their buoys are showing.

While a tide calendar, which charts the rise and the fall of the tide each day, will help a fisherman strategize when and where to haul to an extent, the fisherman must also hold an intimate knowledge of how the tides behave in specific areas. Some sections of the coast are easier to fish during high water, while others are easier to fish at low water. In certain inlets, a fisherman's buoys may only show, or watch, during the high water. The islands and points that pepper Maine's coastline also impact the tide, making it more forceful in some areas and creating eddies in other areas, where the water swirls in the opposite direction of the tide. The result is that the fisherman must dart up and down the coast, making use of small windows of time when the tide is slack and his buoys show.

During the summers that I fished with my father, we spent countless hours scooting up and down the coast of Cutler catching tide slacks. Sometimes we'd arrive at a cove early, to a glassy ocean surface completely void of buoys. Throwing the engine in neutral, we'd "park up," perhaps have some lunch and wait until the tide went slack and pop, pop, pop, the buoys would start to surface. From then on, we'd have about an hour and a half to work the cove until those same buoys started disappearing under the surface of the water.

Down East tides also act differently depending on the time of day and the cycle of the moon. As a rule of thumb in Down East Maine, when low tide occurs at sunset or after dark, that tide will be more extreme. The full moon also brings fuller tides. These forces not only impact how easy it is for a fisherman to haul his gear but also how well his gear will fish. Usually the week before the fullest tides of the month, a lobster fisherman's catch will be better than normal. Peak tides also tend to bring better fishing in shoaler waters. The week following the full tides, the fishing tends to drop off, after which point it will level out. While fishing in peak tides is more challenging, due to the increased velocity of the currents, fishermen will work these tides as best they can to take advantage of the increased lobster catch.[36]

THE SEASONS OF LOBSTER FISHING

SETTING OUT IN THE SPRING

Around early April, depending on the weather, the fishermen get itching to set off their traps. Even though the first few months of fishing will mostly be "changing the water in the traps," it's still important for the fishermen to get out there and mark their territory. Unlike farmers of the land, lobster fishermen don't own specific sections of fishing ground. Legally, each fisherman has as much right to a section of fishing ground as the next. But a fisherman can set precedence by fishing the same section of ground year after year. By getting his traps out early, the fisherman ensures he's able to set his traps in his preferred fishing ground for another season.

If you're in a Down East Maine fishing village during the start of the lobster season, you'll see piles of traps everywhere. The harbor will be filled with boats piled high with traps ready to set. The wharves will be busy with fishermen winching piles of traps off trailers and down onto their boats. The roads will be filled with trucks going back and forth to the harbor, towing trailers piled high with traps. On the sides of the road near the wharves, pickup trucks will be parked up with piles of traps, waiting for a chance to get onto the wharf and unload.

Setting off traps in the spring is an all-hands-on-deck affair. Getting the traps down to the wharf is relatively easy. The more challenging aspect is getting the traps from the wharf down onto the lobster boat. If possible, the fisherman will time trap loading with high tide so that his boat is almost flush with the wharf and traps can be easily handed from wharf to boat. Often,

A wharf loaded with traps, ready for the spring set.

though, a fisherman will have to load his traps when the tide is low and his boat is twelve to sixteen feet lower than the wharf. To help transfer traps to the boat in these instances, a hydraulic winch with a boom, rope and metal hooks is leveraged, and trap loading becomes a three-person affair. One person must operate the winch and boom from the wharf. Another person needs to stand on the trailer, fastening the winch hooks onto the traps so they can be heisted down the side of the wharf. A third person must be stationed down on the lobster boat, receiving the traps. Sometimes a fisherman's wife, son, daughter or neighbor will come down to the wharf and help "hook on"—i.e., be in charge of fastening the winch hooks onto the traps. There is often a wonderful feeling of community at this time of year, as different people stop what they're doing and help out with the loading of the traps.

When loaded, a forty-five-foot boat can carry close to two hundred traps. In addition to traps, the boat will be weighed down with barrels of line, bunches of buoys or trawl balloons and buckets or totes of lobster bait. With a full load, the boat can be carrying an extra six tons of weight, causing the stern to squat down in the water quite a bit. A fully loaded boat can also be quite cramped on board, with just standing room for the fisherman and his sternman in the cabin of the boat. An organized fisherman will have rigged and piled the traps on his boat in such a way that there's little more to do than toss them overboard when he gets to his fishing ground. Usually once the boat is loaded, the fisherman will head right out to set that day, though there are instances when a fisherman will take advantage of an afternoon or evening high tide to load his boat and then moor it overnight before setting in the morning.

Once the traps are set, the fisherman will haul them about once a week, sometimes even less. Again, the lobsters move slowly this time of year. The water is still quite cold—about mid-thirties to low forty degrees—and the lobsters are still coming out of hibernation. During the spring months, a lobster fisherman won't even break even at the end of the week. The price of fuel and bait are just as high as at any other time of year, but the catch tends to be discouragingly low. But the lobsterman knows that summer and fall are just around the corner and today's investments will pay off in the near future. He may even take a loan from the bank to get things going in the spring, confident that he'll be able to pay it back, along with the interest, before the end of the season.

Summer: Shedding Season

From April through July, the fishing tends to be quite slow. Then, by late July, the water starts to warm up significantly—to about the mid-fifty-degree range—and the lobsters start to move inshore. As they move up into the bays and inlets, they are looking for rocks and mud in which to shed their shells.

Because the shell of a lobster is hard and inelastic, it must shed this shell in order to grow. The shedding process can start any time from early June through August, depending mostly on the temperature of the water. During this time, the lobster's shell begins to weaken until it cracks and splits in two up the back, or carapace. At the same time the carapace splits, the tail separates from the rest of the body, and the lobster is able to pull itself free of its old shell.

Beneath the old shell, the lobster has been growing a soft new shell. This shell, and the lobster itself, feels like jelly just after the shed. The remaining old shell is a perfect double of the lobster, down to the claws, legs and even the eyeballs. The newly shed lobster will eat the old shell for its nutrients, which will help the new shell harden. The lobster will then go into hiding for several weeks while the new shell begins to harden.[37]

After a few weeks, the lobster will come out of hiding to seek out food. The shell will still be a very soft, roomy encasing for the lobster. Much of the weight of a "shedder," or newly molted lobster, is water, as disappointed diners who crack open a soft-shell lobster quickly learn. That said, the flavor of a shedder can't be beat.[38]

From a lobsterman's perspective, the shedder season comes on quite slowly. He'll be hauling along and picking out his traps when all of a sudden he'll grab hold of a lobster that's soft to the touch. He'll think to himself, "a ha, the shedders are starting to come on," and he may even announce on the VHF radio that he's caught his first shedder of the season. It may then be another few days or the following week before he catches another shedder. It will take at least several weeks before he's catching shedders in most of his traps.

When the shedders come, the fishing starts to pick up. Shedders are hungry for food, so they are good candidates for trapping. The other thing to remember about shedders is that when they shed they gain about 15 percent in length and 40 percent in weight. Depending on their original size, the process of shedding often pushes them from being a juvenile to a legal-size lobster. Many of the keepers caught by lobstermen in the summer are the same lobsters that were thrown back over the side in the spring. Back then, they didn't make the gauge, but now they're money in the bank.[39]

This very valuable catch is also very fragile and requires special handling. Unlike hard-shell lobsters, shedders must be picked up and moved around more gently. When I was fishing with my father, he'd always let me know if he felt an especially soft shedder while picking out the trap. When I went to band that shedder, I'd make sure to clutch it a bit more loosely and put the bands on slowly and gently. I learned from practice that a swift twisting motion while banding could sometimes tear the claws right off. While my father could still sell a one- or no-clawed lobster, he'd have to market it at the reduced, "cull" price. I would also place a second band on one of the claws to mark the lobster as a shedder so we'd know to handle it carefully when it was time to crate up and sell the catch.

Because of their fragility and the lower ratio of meat to overall weight, shedders are priced and distributed differently than hard-shell lobsters. Where a hard-shell lobster might command a $3.50-per-pound boat price in the summer months, a shedder of the same weight will fetch around $2.50 per pound. The separate pricing structure for hard shells and shedders is established around early July, when the fishermen of western Maine start catching a decent number of shedders. New-shell lobsters also don't fare

Lobster fishing in T-shirt weather. *Courtesy of Gillian Ryan.*

well when transported long distances, so most shedders are either sent to processing plants or served in restaurants along the coast of Maine. Many tourists are willing to make the trek to Vacationland in order to experience the sweet, delicate flavor of a newly shed lobster.

Summer is the most enjoyable season for lobster fishing The weather is fair, the 5:00 a.m. wake-up call is made easier by the beautiful sunrises and the lobster catch has picked up from the depressing volumes of spring. Though fog banks occasionally creep in and disrupt the fishing, a lobsterman will often be able to get through his gear in three or four days, leaving him extra time to putter around or enjoy a long weekend. Though it's almost always chilly out on the Atlantic, on some days a fisherman may even be able to enjoy hauling traps in just a T-shirt.

FALL FISHING: BOOM TIME

By around Labor Day, up to 90 percent of a Down East lobster fisherman's catch will be composed of hardened-up shedders. These lobsters are now extremely active. They are beginning to migrate into deeper waters offshore, and as they move across the ocean bottom, many will end up in fishermen's traps. This is the boom time of lobster fishing, with catches peaking in the months of September through November.

In order to maximize catches during the fall months, lobster fishermen do a lot of trap shifting. Those fishing traps in depths of ten to fifteen fathoms along the shore will be moved out to thirty fathoms or deeper. Many fishermen will also plant traps on muddy bottom. The goal with all this gear shifting is to follow the migration routes of the lobsters as they move across the bottom.

A fisherman wants to stay just ahead of the lobsters so that when they migrate through an area they end up in his traps. Fishermen often call this "catching a run" of lobsters. As lobsters don't move at a steady pace over the bottom, it's a challenge to anticipate in which areas they'll be and for how long they'll stay. Similar to a smart stock investor, a skilled fisherman will spread his bets. He'll place a healthy percentage of his traps along bottom, which is a sure bet for catching lobsters. He'll also sprinkle a small portion of traps along areas of bottom that are not yet fishing well. If one of these outlying traps starts fishing well, he'll be the first to know and can quickly shift his gear to catch the run of lobsters.

A lobster fisherman doesn't just want to stay ahead of the lobsters; he wants to stay ahead of the other fishermen as well, and spreading his bets is one way to do so. If ten fishermen are occupying a section of ground when the lobsters migrate through, each fisherman will likely get 10 percent of the keepers from that run. On the other hand, if the fisherman is able to stealthily catch a run that other fishermen have not discovered, he'll get a far larger percentage of lobsters as they migrate through the area.

One other way a fisherman can maximize his catches during the bustling fall fishing months is by hauling back frequently during the week. There are so many lobsters moving through a fisherman's traps in the fall that frequently refreshing the bait and clearing out the traps' bedrooms to make way for more lobsters is a smart investment.

Fall fishing is the busiest time of the year for a lobsterman. He and his sternman will leave dock before sunrise and return after sunset. Most fishermen are out of the harbor no later than 5:00 a.m., navigating their way through the darkness with spotlights mounted to the roofs of their boats. When the sun rises later in the morning, it brings better visibility but little warmth. Layers of fleece and flannel are worn to help fend off the chilly Atlantic air. Unless the weather is especially foul, the fishermen won't return to the harbor until 5:00 p.m. or later. Even when lobstermen manage to get back to port a bit earlier on a fall day, they often must spend several hours

Above: My father, his sternman, Lee Hildreth, and me fast at work on a chilly fall day. *Courtesy of Anthony Oragano.*

Left: A picking box full of lobster on a busy fall day.

Cutler fisherman Stephen Cates after a long day of fall fishing. Steven started lobster fishing at the age of five and has lobster fished his whole life. *Courtesy of Anthony Oragano.*

banding and crating up their catch before being able to head home. The volume of lobster is so great this time of year that it's simply not possible to band most of the catch until the end of the day. While working Saturdays is not uncommon for a fisherman in any part of the season, during the fall, Saturday fishing is especially frequent. Sometimes a fisherman needs to haul on Saturday because poor weather has prevented a day of hauling earlier in the week. Other times, a fisherman will want to squeeze in an extra day to maximize his weekly catch.

THE WINTER MONTHS: MAKE AND MEND

By early to mid-December, the weather starts to get pretty mean in Down East Maine. The winds become stronger, and stormy days are so frequent that it becomes difficult for the fishermen to get out and tend their traps. At the same time, cooling waters have caused the lobsters to slow down or "drop off." At some point in mid- to late December, the lobsters become so few and the weather becomes so tempestuous that the fishermen decide to end the season.

Winching a load of traps off the boat at the end of the fishing season. *Courtesy of Anthony Oragano.*

Unloading traps at the end of the season.

Rush hour in Cutler, Maine. A fisherman takes home his traps at the end of the season. *Courtesy of Bill Kitchen.*

At the end of the season, the harbor is once again filled with boats piled high with lobster traps ready to unload. The wharves are busy with fishermen winching traps up from their boats and onto trailers, and the roads are filled with trucks going back and forth to the harbor with trap trailers.

Up to 70 percent of Down East fishermen take up all their traps in the winter months. Once the traps are up, some fishermen haul their boats ashore for winter as well. While Maine harbors don't freeze over in the winter, some fishermen have older or smaller lobster boats that they want to protect from the winter elements. Storing the boat ashore offers these fishermen peace of mind, as well as the opportunity to perform needed repairs. Those fishermen who leave their boats in the water through the winter, including my father and brother, will make sure the mooring line is secure and the engine has had proper maintenance.[40]

Once his traps and boat are secure for the winter, the fisherman will likely purchase some new gear to replace traps that have been damaged or lost in the previous season. Often, up to 10 percent of a fisherman's gang will need to be replaced each year due to shrinkage. The fisherman will also overhaul his current gear, a process that can take weeks to months, depending on the level of focus to the task at hand. All trap elements are inspected in the overhauling process; the trapdoors, heads, bridles and guy strings are checked for robustness. All the trapline is overhauled as well to check for wear and tear. Biodegradable hog rings in the escape vents are replaced, and new trap tags are snapped into place. The work is monotonous, and there's no paycheck at the end of the day, but there is some satisfaction in seeing gear shift from an undone pile to the done pile.

After trap repairs are complete, the fisherman will move on to repainting his buoys. Crusty sea growth is first scraped off the buoys, and then one or several fresh coats of paint are applied in the fisherman's unique buoy color pattern. In foul weather, the trap repairs and buoy painting will take place in a shed or garage. In the spring, such activity often takes place outside, in a trap lot or down on a wharf, as the fishermen enjoy the fresh air while battling the infamous Maine black flies (sometimes jokingly referred to as the Maine state bird).

Buoy painting was one of the first assignments I was given in the world of lobster fishing. Each spring, starting around the age of eight, I would work my way through seemingly endless piles of buoys, scraping them, applying three

Trawl balloons collecting snow in the winter. *Courtesy of Bill Kitchen.*

Boats at rest in Cutler Harbor on a cold winter's day. *Courtesy of Bill Kitchen.*

stripes of fresh red paint and then hanging them on our family clothesline to dry. I continued painting my father's buoys well through college and still enjoy the process of painting buoys when I can find the time.

THE MAINE LOBSTER MARKET

Most fishermen in Down East Maine "belong to" or "fish for" a specific lobster dealer or cooperative. The fisherman will sell his catch exclusively to this dealer or cooperative for a specified boat price. In exchange, the lobsterman is given a guaranteed market for his lobsters. The dealer or cooperative will store the lobsters for several days to several months before selling them to another wholesaler or a retailer. Eventually, much of Maine's hard-shell lobsters end up on the Boston market or in restaurants along the coast of Maine. Each time the lobster changes hands, the price goes up. A lobster sold by a lobster fisherman for a boat price of three dollers might cost twenty dollars when it's put on your dinner plate.

Given the vast markup in lobster prices between the first and final sale, it is often asked why lobster fishermen don't take their lobsters farther than from trap to dock, cutting out the middlemen by getting involved in distribution and marketing. The answer becomes clear when one understands the complexities of the market, the relative frailty of lobsters and the powerful social and cultural factors at play.

From a market perspective, the price of lobster, like oil, is in constant fluctuation. A fisherman can leave to haul in the morning expecting a particular price for his catch only to discover that the boat price has dropped thirty cents when he lands his catch that evening. Though there is a broad cyclical pattern in the lobster market, with prices peaking in the winter months, short-term prices are impossible to predict.[41]

While the lobster fisherman could, in theory, time the sale of his catch to meet rises in market prices, he is dealing with an extremely perishable product. If a quantity of lobsters is confined in a close space for any period

The Maine Lobster Market

Right: Weighing up lobsters.

Below: My father and brother getting ready to sell their catch. Each gray crate is filled to the brim with lobsters. *Courtesy of Bill Kitchen.*

of time, a certain percentage of those lobsters will perish. The longer the lobsters are held in lobster crates or lobster cars, the greater the shrinkage. While at some points the market price for lobster will be high, at other points the market is almost nonexistent. It is usually advantageous for the fisherman to unload his catch as quickly as possible versus holding out for a golden moment. By committing to a dealer, the fisherman has a guaranteed market for his highly perishable product, regardless of the market conditions.

It could still be asked why lobster fishermen don't practice vertical integration, buying up a fleet of boats to fish for lobsters, along with several dealerships and lobster pounds to help properly store and move their product. Such firms do not exist due to powerful social and cultural factors. In essence, any entrepreneur who tried such a move would immediately run afoul of the deeply rooted territorial system that exists in lobster fishing.[42]

The Lobster Market throughout the Season

With lobster fishing, for the most part, catch and price are inversely related. The price comes to its peak during February and March, when catches are at their annual low. In the fall, when catches reach their peak, fishermen receive the lowest boat prices of the year.

Some fishermen play the market to a small extent by storing up lobsters in cars or fishing for a cooperative and then selling their catches in bulk during a price spike. For the most part, however, fishermen base their fishing efforts on the availability of lobster and the seasonal weather conditions instead of the market price. In the winter months, the weather is so poor and the lobsters so few that a day on the boat often won't pay off, despite the high market value of lobster. The scarcity of lobster, in part, fuels these annual high market prices. In the late summer and early fall, lobsters are so plentiful that the big catches more than make up for the lower market prices.

In general, the market price of lobster peaks in the winter, when the water is cold and the fishing drops off. One price peak usually comes just before New Year's Eve, as consumers celebrate the end of the year with fine dining. The price then remains quite high until the spring, when the water temperature warms, lobsters become more active and fishermen begin tending their gear. Price is usually deflated through late spring and early summer due to low demand. It picks up again in July and August, due, in part, to the influx of tourists to Maine. Fall prices fluctuate depending on the volume of the catch. Sadly for fishermen, seasons that bring record-breaking catches are also often met with record-breaking low boat prices.[43]

THE PERILS OF
LOBSTER FISHING

You gotta respect the ocean, because sometimes you don't get a second chance.
—Norbert Lemieux

ACCIDENTS AND DEATHS

Maine lobster fishing is mainly a day fishery. It is less perilous than other forms of commercial fishing, which see men and women sent off to work Georges Bank for months on end. When it comes to navigating the open seas, however, there is always an element of risk. One mistake could be your last. On average, a Maine lobsterman dies at work about every other year. Many more are injured on the job. Broken bones, missing fingers and bad backs are common afflictions of a life spent on the ocean.[44]

The Gulf of Maine is the perfect, nurturing environment for *Homarus americanus*. It is unforgiving to any fisherman who is plunged into its chilly depths. The ocean Down East is notorious for its frigidity, with temperatures ranging between thirty-eight and sixty-two degrees Fahrenheit. Hypothermia can set in within minutes. The expected survival time of a person in sixty-two-degree water ranges from two to forty hours. In thirty-eight-degree water, survival time can be an hour or less.[45]

The temperature of the water would be irrelevant for some fishermen if tossed overboard. A fair few don't know how to swim in the first place. Those who can swim would struggle against the weight of their heavy boots and oil gear to stay afloat. While it is now a legal requirement for all lobster fishing boats to be equipped with insulated survival suits, which can keep a fisherman afloat and protected from hypothermia, things often go wrong so quickly on a boat that donning a survival suit isn't an option.

Getting your hand or foot caught in moving line on the boat is one of the quickest ways to get into trouble. When setting a trap, the line zips down the length of the boat's floor before exiting over the stern, propelled by the force of a heavy, sinking trap and forward-moving boat. In a flash, that line can wrap around a rubber boot just like a lasso and reel you into the icy water. About 25 percent of the fishermen I interviewed admitted to getting their hand or foot caught in the line at one point while out at traps. Thankfully, none of these fishermen was hauled overboard in the process.

My father is one such fisherman lucky enough to be here to tell the tale of getting caught in the line. While in the process of setting a trap, he reached his hand down to push back the coil of moving rope. His eyes were focused on the next buoy ahead, and his intention was to clear the way for that trap. Before he knew it, the line had jumped up and looped around his right wrist. The strain of the rope was immediate. With the boat still in gear, he ran and dove under the stern, bracing his feet against the decks to fortify against the pull of the line. The pressure was intense, but he was able to wrestle himself enough slack on the line to pull his hand free before being hauled overboard.

While getting caught in a line exiting the boat is one of the biggest perils of lobster fishing, hauling line into the boat is a risky endeavour as well. The hydraulic hauler system, with its rapidly rotating disks, can easily catch hands or fingers. Of the sixty fishermen I spoke to for this book, 20 percent had broken or strained a finger, hand or wrist in the pot hauler while fishing.

The pot hauler also creates an incredible tension on the line being pulled into the boat, which can lead to yet further injuries. Sometimes the heavy metal snatch block pulley system lets go under the strain, swoops around and wallops the fisherman in the head. Other times, a fisherman can get injured while trying to clear tightly tangled line as it enters the boat. A line entanglement cost one Cutler fisherman his thumb. The fisherman was trying to clear a labyrinth of tangled line while hauling a trawl. He had stopped the pot hauler and was working on the tangle when the strain of the line became so great that the rope slipped out of the hauler, catching his gloved thumb in the process. Within seconds, the strain tore his thumb right out of its socket. My brother recovered the bloody thumb the following day when hauling back the line.

As with most lobster-fishing injuries, the aftermath of the thumb incident played out over the VHF. Thankfully, the fisherman had two sternmen aboard his boat. One of these sternmen worked with the injured fisherman to try to stem his blood loss. The other sternman gunned the boat for the harbor while radioing in for help ashore. My mother was the first to pick up the VHF. She called for an ambulance and then, over the VHF, relayed advice she received from medical staff on the phone to the fishermen on the boat. Upon reaching the harbor, the thumbless fisherman was airflighted to a nearby medical facility. While he never regained his finger, he has recovered well and continues to lobster fish full time.

The day my father got his hand caught in the line and the day of the thumb incident were relatively calm days on the water. When the weather is foul and the seas are choppy, the risks are even greater. With more than forty knots of wind, things happen fast. A fisherman with good marine common sense will know how much his boat can handle and when to head in. He'll also have good knowledge of all the partially exposed ledges in the area he fishes. A sea will cut down the amount of water covering those ledges, and the surge around the ledges will also be stronger when conditions are stormy. Many ledges along the coast of Maine have been named after fishermen who've gone ashore on them in stormy weather. In Cutler, we have Zwinkel's Ledge, a small, rocky outcrop detached from the mainland, which only shows at some points of the tide. It's named after Erich Zwinkel, who crashed his boat on the ledge during a stormy day in the mid-1900s. After the crash, the boat sank, and Zwinkel remained stranded on the ledge. He was eventually rescued by my grandfather and another man from my town. Zwinkel's first words to my grandfather upon being saved were, "I knew you'd come for me, Glen."

There are many other stories that don't have such a happy ending. Every harbor seems to have at least one tale of a life lost in the sea. To my knowledge, two have drowned from Cutler Harbor. In 2010, a Lost Fisherman's Memorial was erected in Lubec, Maine, as a lasting tribute to all those lives lost at sea in the Maine and New Brunswick area.

Hurricanes and Storms

A hurricane running up the eastern seaboard in the fall is one of a lobster fisherman's worst fears. The high winds and surging seas can really do some damage to the fishermen's traps, boats and wharves. If the storm hits during the high tides, it's especially distressing. Once the wind gets to sixty or seventy knots, fishermen start to worry about their boats, even in the safety of the harbor.

To prepare for a hurricane, any lobster fishermen fishing gear in shoal water or in close to the shore will take that gear, usually fifteen- and twenty-fathom traps, and either place it in protected coves or take it right ashore. If you've ever walked along the shores of Maine and spied a mangled-up wire trap or a buoy stuck way up high in some ledges, it was likely the result of a trap being left too close to shore during a hurricane.

In the harbor, skiffs will be hauled out of the water and up onto floats, where they will be flipped over so as not to be swamped by the rain. The scales used to weigh lobsters will be moved off the floats and stowed in a safer area. Any extra traps, buoys and rope stored on wharves will be removed or secured so as not to be blown about. In extreme cases, when a strong hurricane strikes during a high tide, fishermen have been known to park their trucks down on the wharves to keep the strong seas from lifting off the wharves' planks.

Boat moorings will also be prepared for hurricane-force winds. The boat owner will ensure his mooring line is in good shape and skin off any marine growth that could put an extra strain on the line in a surging sea. Some fishermen will also double up the lines on their moorings. If a boat breaks loose from its mooring in a hurricane, it means bad news not only to the boat owner but also to the owners of all the surrounding boats into which it can smash. In the throes of a hurricane, most fishermen will go down to the harbor every several hours through the day and night, checking to see if all the boats are okay.

In the spring of 2011, a tropical storm swept through Down East Maine, bringing with it winds of over sixty knots. Before the storm had passed, two boats and one lobster car in our harbor had broken loose from their moorings. The first boat snapped its mooring line, drifted across the harbor and landed on some ledges along the shore. Strong winds then kept it pinned against the ledges while big seas lifted the boat up and down, bashing it against the rocks. A group of fishermen went down to the shore to try to tie the boat off as best they could, only to witness a second boat breaking free of its mooring. This second boat was rescued before it reached the shore by a brave fisherman who leapt into his outboard tender, raced out to the boat, jumped aboard and got her running in time to steer her to safety. Meanwhile, a larger group of local fishermen had gathered and set to work on getting the first boat off the ledges. In the end, it took fifteen fishermen, three skidders, an excavator, a hydraulic jack and a boat trailer to liberate the vessel.

THE UNWRITTEN RULES OF LOBSTER FISHING

The Territorial System

The Maine lobster-fishing industry is highly territorial. Lobstermen have long lived under unofficial rules that dictate who can set their traps where along the rugged coast. These rules are legally unenforceable but important and usually accepted. In fact, many fishermen believe self-governance is the best way to regulate the industry and keep the lobster stock healthy. If the record catches in the last thirty years are any indication, these lobster fishermen might well be on to something. To protect lucrative fishing grounds, some fishermen even resort to frontier justice, cutting lobster trap lines, ramming boats and occasionally drawing arms.

In order to comprehend how such violence can come to pass, one has to understand the dynamics of these small-town fishing communities. First of all, the people who fish from the same harbor are some of the most important people in a fisherman's life. He relies on his fishing community in times of need, from getting help with loading traps to an offshore rescue if his boat breaks down. If he becomes injured, local fishermen will often haul his traps and give him the proceeds until he recovers. A lobsterman also must be able to trust his fellow fishermen. You can't put a padlock on a lobster trap, and it's practically impossible for officials to police the seas ensuring that one fisherman does not haul another's gear.

In part because trust and accountability are so critical to fishing communities, many Maine fishing villages have an unwritten rule that you can't catch lobster from that village's fishing grounds unless you live in the actual town and are part of the community. It's much easier to be dishonest with a fellow fisherman if you don't have to bump into him at the local store or village church. Being part of the fabric of the fishing village is seen to make a fisherman more accountable for his behavior on and off the water.

Perhaps more importantly, people "from away" who encroach on local fishing grounds are also seen to be taking resources away from that community. The Gulf of Maine is incredibly fruitful, but there is still a finite amount of lobster that can be harvested from any section of fishing ground. If an out-of-towner encroaches on a village's section of ground, it leaves less harvest for local citizens. The concept that local resources, including jobs, land and fishing grounds, should be reserved for local people is deeply rooted in the Down East Maine psyche. Additionally, many of the local fishermen in one's community are family members. Up to 85 percent of those I interviewed for this book are part of a generation of fishermen. Lobster-fishing fathers are territorial in part because they want to protect the resource for their sons and grandsons. These sons and grandsons usually grow up wanting to become lobster fishermen. They have few other viable jobs to turn to in Down East Maine regardless. If someone sweeps in from away, with no connection to lobster fishing, and tries to fish the grounds, it goes down like a lead balloon.[46]

The ocean floor and the lobsters that live there are, by Maine law, public resources. Any fisherman with a lobstering permit has the legal right to fish in Cutler or pretty much anywhere along the coast. Yet those entering the lobster-fishing industry quickly learn about the invisible lines that carve up the fishing grounds along Maine's rugged shoreline. There is no written document specifying the GPS coordinates of these boundaries, but the details are known and mostly observed by the local people who fish that area. Points, coves and islands are often used as unofficial markers. Cutler's fishing grounds extend from Bog Cove at the eastern end to Cross Island at the western. Cutler fishermen know not to fish to the east of Bog Cove and, at the same time, will not tolerate fishermen from another territory who place their traps to the west of the Bog Cove boundary line.

At times, there are even subdivisions of fishing territories within a local fishing community. During the mid-1900s, my town had two commercial lobster-fishing wharves. Those fishermen who fished for the first wharf, owned by my grandfather, fished to the east of Cutler Harbor. Those who fished for the second wharf, owned by my great-uncle Neil, fished to the west of Cutler Harbor. No one in Cutler Harbor could tell you exactly why that was. They just knew to turn east or west depending on the wharf for which they fished. While today the "east or west" rule doesn't always hold true, the Bog Cove and Cross Island boundaries are still fiercely protected by Cutler fishermen. Outsiders who try to encroach on our fishing grounds will likely be met with vigilante justice, an approach practiced by fishermen throughout the state.

Trap Wars

The methods used by lobster fishermen to protect their fishing grounds have gained a level of notoriety for their stealth and the occasional act of violence. While stories of ramming boats and firing weapons are not unfounded, fishing feuds are often resolved more quietly. Often referred to as "trap wars," these skirmishes are as old as lobster fishing itself, and the codes of conduct that govern these wars, like the fishing boundaries, are unwritten yet known and practiced by those in the fishing community.

If an outsider encroaches on a community's fishing grounds, the violater is first given a "friendly warning." While it may be a verbal threat, more often it is a cloak-and-dagger molestation of the fisherman's gear. Common practices include hauling a bunch of the fisherman's traps and cutting out the heads or leaving the trap doors open. Other times, a knife is stuck in the side of a Styrofoam buoy or two half hitches of rope are tied around the buoy's spindle. When the wrongdoer next goes to haul and discovers the "signs," he should know to remove his traps from that area immediately. If he doesn't, the next time he goes to tend his gear it won't be there.[47]

Interlopers who don't heed the covert yet clear warnings of the lobster-fishing territory system will most often have their traps cut off. Other fishermen from the town will "take a knife to them," cutting the line between the buoy and the trap so the fisherman's gear is forever lost at sea. Again, this trap cutting usually happens on the quiet so that the wrongdoer is not able to point a finger or go to the law. Seeking justice would be a futile process regardless. While a judge could perhaps rule that the interloper be remunerated for the financial loss of his gear, the man would never be welcome in the town's fishing territory.

Beyond trap cutting, there are instances when retaliation is of a more extreme and violent nature. Fishermen have sometimes been known to ram into a violator's lobster boat at full speed. This was the case in Portland Harbor one year, when a crew rammed its boat into another vessel, jumped aboard it pirate style and scuffled with the other crew before getting tossed overboard. One summer, two boats were sunk in Owl's Head, Maine. The following summer, eleven lobster boats were cut loose from their moorings in Stonington, and a stack of traps was set afire in a lobsterman's yard in Cushing.[48]

In one infamous case from 2000, two lobstermen duked it out with a pitchfork and a fish gaff. According to the case history, David McMahan and Gerald Brown had professional differences for quite some time. After discovering an obscene drawing on one of his buoys, Brown confronted McMahan on his dock. A verbal exchange ensued, rife with salty insults. The verbal assault then turned physical, with McMahan using a gaff to clobber Brown and, according to McMahan, Brown using a pitchfork to jab at McMahan. After the confrontation, Brown, bleeding from the head, sold his lobsters, moored his boat and then drove himself to the hospital.[49]

But perhaps the most famous lobster trap war stories of all hail from the small island of Matinicus, Maine. Over twenty miles from mainland, Matinicus is by far the most isolated of Maine's fifteen year-round island communities. With no paved roads or amenities and a ferry that comes just once a month in the wintertime, the island attracts rugged individuals. For the most part, its hardworking citizens are simply renegades of conventional society. At times, however, they have been renegades of the law as well.

One year, a Matinicus fisherman fired a shotgun across the bow of another's boat when it crossed his wake at high speed. Another year, smelly lobster bait was dumped into someone's gasoline tank. In the summer of 2009, the tiny island with a population of under forty in the winter months attracted national media attention after one fisherman shot another in the neck in the midst of a heated trap war.

The trap war kicked off when a man from Wheeler's Bay, named Alan Miller, decided to set gear in Matinicus's fishing territory. Though Allen's wife and stepfather, Vance Bunker, were residents of Matinicus, many on the island still felt Alan did not have the right to lobster fish their grounds. Soon, traps were getting cut and accusations were being made. Early one July morning, all hell broke loose. First, Vance's boat was boarded by Matinicus fisherman Christopher Young, who wrestled Vance to the ground, threatening to kill him until Bunker emptied a can of pepper spray in his face. Later that morning, there was a boat chase, and threats were made over the VHF. By the afternoon, things had culminated in a face-off on a Matinicus pier. Guns were drawn, and Young was shot in the neck. The aftermath saw Bunker arrested and Young helicoptered to a hospital for surgery. Bunker was later acquitted of the crime after testifying that he had been acting in self-defense.[50]

Though stories like the above make for good reading, I must note that violence is an extremely small part of the daily lives of Down East Maine lobster fishermen. Even Matinicus, which has garnered a reputation for outlaw behavior, is largely just an island of hardworking fishermen with the strength of spirit to survive an isolated way of living.

OTHER UNWRITTEN RULES

There are a host of other unwritten rules in the game of lobster fishing that won't incite violence but may get you a firm talking-to from one of the other fishermen. In the harbor, one of the minor no-nos is not respecting another fisherman's equipment. Often, one fisherman will borrow the skiff of another. When he returns it, it is common courtesy to tie up the skiff using the same knot as used when he took the skiff off the dock.

At the same time, a fisherman should take care not to block access to key areas in and around the wharf. For example, he shouldn't leave his truck parked down on the wharf if another fisherman needs to use the wharf to load traps or bait. He should also be mindful of not spending too much time tying his boat up to the wharf at the end of the day, creating a traffic jam for the other fishermen.

Within the harbor, fishermen must take care not to travel at unnecessary speeds, creating big wakes for other fishermen moored up or traveling around in their skiffs. Many harbors actually have speed limits, though they're never officially enforced. Rather, if someone happens to roar down the harbor and it disturbs another fisherman, he'll usually get a talking-to by the said fisherman or the harbormaster.

Finally, a fisherman should always take care of his mooring so that there's no chance of his boat breaking loose in a storm and damaging the other boats in the harbor. Sometimes accidents can't be avoided, though moorings that are not well maintained are often those from which a boat breaks loose. Fishermen tend to be less inclined to rush to the aid of a drifting boat if the situation could have been avoided in the first place.

For example, once a boat broke loose in Cutler Harbor during the highest tides of the spring, drifted across the harbor and ground out on a bunch of ledges in an upright position. The owner of the boat hadn't been doing proper maintenance on his mooring. The fellow was also a summer resident, and the damaged boat was his pleasure craft. When the storm was over, the boat owner went around to the people in town asking for help to get his boat off the ledges. No one obliged. His boat stayed stuck on those ledges in the harbor all summer until the next set of high tides came in the fall.

Another no-no is unnecessarily tangling with another fisherman's gear. Tangles happen all the time on the water. Tides, swells and fishing close together cause lines to overlap, and this creates some awful snarls. If the snarl is perceived as an accident, the fisherman will be frustrated but courteous when untangling his trapline from that of the other fisherman. If line needs to be cut to get the traps free of each other, the fisherman will usually cut his own line versus the line of the other fisherman as a courtesy. Less consideration will be given to the other fisherman if the tangle is the result of careless fishing. If line needs to be cut to get the gear free, the fisherman will often choose to cut the line of the careless fisherman.

TRAP MOLESTING

Beyond interloping in another town's fishing territory, there is one other act that can trigger a trap war: hauling another fisherman's gear and poaching his lobsters. Fishermen obviously don't padlock their traps, so, if inclined, it's quite easy for one fisherman to haul the traps of another and take his lobsters. Who would know if no one is looking?

Yet a fisherman almost always knows if another person has hauled his trap. First of all, the trap is rarely reset exactly where the owner originally placed it. A good fisherman has a strong understanding of where each trap sets in the water. Slight shifts, which can't be accounted for by tides or storms, are mentally noted. Additionally, a good fisherman can often tell just by the way the trap has been set that another person has hauled it. Sometimes the bait bag is hanging oddly in the trap or the trap is not fishing as well as would be expected for the area and the season.

If a fisherman suspects his gear is being hauled in a certain area, he'll first have a quiet word with others who fish the same area to see if they are having a similar experience. Often, he and the other fishermen will already have a sense of who the culprit might be. A smart fisherman always keeps an eye on the fishing patterns of those around him. If one fisherman consistently avoids hauling his gear at the same times or in the same areas as other fishermen, he will draw suspicion. Further suspicion is drawn if a fisherman spends an unusually long period of time fishing in an area where he has a limited number of traps.

Quiet conversations between fishermen are usually followed by more covert observations of the situation. Are the suspected culprit's catches unusually high versus those of the other fishermen? Are the other fishermen's traps coming up bone dry when the fishing should be good? All of this evidence will help the fishermen decide if they believe their gear is being molested. If the fishermen are pretty much certain another person is hauling their traps, they'll follow the same path of trap war retribution as previously described.

SUPERSTITIONS

Fishermen are known for their superstitions, and lobster fishermen are no different than the rest of the lot. Many lobster fishermen won't paint their boats blue, believing that a blue boat brings bad luck. Others won't go aboard a boat if one of the hatches is open, as an open hatch is considered bad luck as well. Various fishermen like to stick to the same routines when they start their days, doing the same things in the same order when they board their boats.

Many fishermen also refuse to start a major project on a Friday, including starting a fishing season. If the weather is poor except on Fridays for several weeks in the spring, most fishermen would rather lose those weeks of fishing than set off their gear on an unlucky Friday. My father adheres to this superstition yet hates delaying his fishing. When it comes to setting gear in the spring, my father will pay close attention to the weather forecast. If it looks to be unsuitable for setting off Monday through Thursday but good on Friday, my father will still venture out on Thursday, taking a few fifteen-fathom traps, which he'll set just outside the harbor. This small, symbolic trap setting ensures that he doesn't technically start his season on a Friday.

THE WRITTEN RULES
OF LOBSTER FISHING

The first thing to note about written rules is that lobster fishermen don't tend to like them. This is not to say that lobstermen don't believe in conserving their resource. Rather, most Maine fishermen believe that the lobster industry functions most effectively when it is self-regulated. Unwritten rules, such as the territorial system and trap war retribution, play a key role in keeping resource exploitation at bay. Challenging seasons, when lobster stocks are down or expenses are up, help with natural selection, driving out those fishermen just looking to make a quick buck. Additionally, many of the written rules are conceived and imposed by politicians and scientists who have a limited view of their actual impact to the lobster stock and the industry itself. In fact, 70 percent of surveyed fishermen state that their biggest concern with their job is government regulation.[51]

Most fishermen have a deep distrust of "the feds." One fisherman from my survey spoke of their "crazy laws." Though lobstermen respect the objective to preserve lobster resources for the good of future generations, most don't trust government's and scientists' views of the industry or their strategies for conservation. For over three decades, federal fisheries managers have issued doom-and-gloom warnings that the lobster stock is overfished and on the verge of collapse. Those same three decades have given way to year after year of record-breaking lobster catches. Clearly, at least some of the federal fisheries' industry observations have been flawed. Suggested government regulations that might make sense on paper can often have adverse effects in the real world. For example, the Maine lobster trap quota, imposed to help limit the harvest, has become a benchmark for some Down East fishermen, prompting them to increase the size of their gangs to the current eight-hundred-trap limit. Other rules don't seem to acknowledge the complexity and unpredictability of the ocean environment.

Yet there is a series of rules that lobster fishermen respect and support. These regulations focus not on restricting fishermen but rather on protecting lobsters at critical points in their life cycle, such as when they reproduce. The fishermen themselves lobbied for several of these rules. All are now regarded as critical to the success of the Maine lobster fishery today.

MAINE LOBSTER FISHING'S FOUR CRITICAL CONSERVATION LAWS

1. Young lobsters, called juveniles, cannot be harvested. Until they reach a size of over 3.25 inches on the carapace, they must be returned to the ocean. A metal gauge is used to check the carapace size. An average lobster in Maine waters will live and grow for about seven years before it is of harvestable size.[52]

2. Large lobsters, more than five inches on the carapace, cannot be harvested. These lobsters are considered "forever wild" and must be returned to the ocean when caught. As with small lobsters, a metal gauge is used to check the carapace size of large lobsters.

3. Female lobsters that are pregnant (egg-bearing) or marked with a special, man-made notch in their inner right flipper cannot be harvested. They must be returned to the wild. Months or years down the line, if these female lobsters are no longer bearing eggs or have outshed their V-notches, it may be possible to harvest them.

4. All lobsters must be caught in traps—no dragging or diving is allowed. The traps must include escape vents for undersize lobsters, as well as biodegradable escape hatches to free lobsters in lost traps.

The minimum-size law and the law banning the taking of egg-bearing females were the first regulations put in place by the government. Dating from the late 1800s, these laws are relatively basic and make sense to the casual observer. More interesting and noteworthy are Maine's unique maximum-size law and V-notch program. These laws have come about largely due to the fishermen and reflect the lobstermen's strong commitment to the conservation of the industry.[53]

The V-Notch Program

The V-notch program was started in the early 1900s as a more rigorous way to protect egg-bearing female lobsters. Under government instructions, wardens began marking the inner-right flipper of a set number of egg-bearing female lobsters each year. This mark, a V-shaped notch, meant that the female lobster could not be legally sold in the state of Maine—whether she was carrying eggs or not. She was free to roam the ocean and reproduce for several more seasons until, through the process of molting, she outshed the V-notch. The fisherman's obligation was to check each female lobster caught for a notch and return any notched lobsters to the ocean.

Maine lobstermen, however, soon began going above the call of duty to return V-notched lobsters to the sea. They began notching lobsters themselves. It was easy for fishermen to see the logic in protecting reproductive females. Each time they notched a fertile lobster, it was like placing an investment in the future of the industry. The V-notch also assured a fisherman that each egged-out lobster he threw back in the sea could not be caught and sold by another fisherman once that lobster had released her eggs. From the mid-

A V-notched lobster tail.

1900s onward, V-notching female lobsters was a completely voluntary yet widely practiced exercise by fishermen up and down the coast of Maine. In fact, in the late 1900s, when biologists argued that the V-notch law should be abolished, lobster fishermen fought vehemently to protect the practice. Today, most researchers, along with the government and fishermen, agree that the V-notch program is one of the key reasons why the Maine lobster industry is achieving record-high catches. Now, at certain points in the season, a fisherman will throw back a V-notched lobster from every trap he fishes. Recently, legislature even passed a law making it mandatory for fishermen to V-notch all the egged-out females they catch. While this law will be pretty much impossible to enforce, unless the government plans to install a surveillance camera on every fisherman's boat, it's highly likely that the fishermen will continue adhering to the practice of V-notching due to their commitment to the future of the industry.[54]

The Maximum-Size Law

In addition to the V-notch program, the maximum-size law is another Maine lobster-fishing conservation practice defended by the fishermen themselves when government scientists pushed to have it abolished in the late 1900s. At first glance, it would seem illogical for lobstermen to fight for a regulation that would keep them from selling the largest, most profitable part of their catch. An oversized lobster with a weight of five pounds could easily sell for twenty-five dollars in the right market conditions. At certain parts of the season, a full-time Down East Maine fisherman can catch as many as twenty lobsters a day that weigh in excess of five pounds. Some of these lobsters will way upward of ten pounds. Yet scientific research and fishermen's observations have shown that the long-term value of conserving Maine's largest lobsters is far greater than selling them for short-term profits.[55]

Unlike humans, as female lobsters get older they become more reproductive. Rather than becoming barren, larger, older female lobsters produce eggs more often and in much greater quantities than their younger counterparts. It has even been estimated that it would take over twenty-seven small one-pound lobsters to match the reproductive powers of a five-pound female over a five-year period. While some of these large females are protected through the V-notch program, this is only half the battle. It also takes a massive male lobster to woo and impregnate these jumbo females. In other words, it takes two large lobsters to tango. The maximum legal lobster size ensures that all lobsters with a carapace over five inches in length are free to roam the ocean and mate with one another to their hearts' content. It's quite a pleasant retirement for a sea creature—and a retirement that is only possible due to the dedication of fishermen. [56]

Lobster Trapping

The size laws and the protection of a large portion of female lobsters are the foundations of Maine's lobster conservation program. These laws ensure that for every lobster that comes to the market, about three go back into the water for conservation. Adherence to these regulations has helped the Maine lobster industry become a model of sustainable fishing for the rest of the United States.[57]

Beyond these conservation measures, the practice of trapping lobsters has also helped the industry deliver record-breaking catches for the last thirty years. In fact, as the number of traps fished has increased along the coast of Maine, so has the lobster stock. It might seem counterintuitive that fishing more traps more frequently helps bolster the lobster resource. Yet in reality, each time a trap is fished, a bunch of lobsters get a free meal. Shorts, bulls, eggers and V-notchers are returned to the sea with full bellies after each haul. The trap itself is sent back to the ocean floor with a full bag of bait, upon which more freeloaders can feast. Small openings in the trap, called escape vents, allow many of the smaller lobsters to pass freely through the trap between hauls. The number of trapped lobsters a fisherman sees is only a portion of the crustaceans that have feasted on his bait. In the event that the trap's line is parted and the trap becomes lost on the ocean floor, one wall of the trap is designed as biodegradable, allowing larger, trapped lobsters to break free.

LOOKING TO THE FUTURE

Trap after trap, day after day, season after season, a lobster fisherman observes the habits of the species upon which he makes his living. By the time they are in their forties, most lobstermen have spent over three decades studying the patterns of their prey as they chase lobsters across the ocean floor. This knowledge is often supplemented by a wealth of insights passed down from previous generations of fishermen in their families. Yet because lobstermen tend not to have advanced degrees or scientific data to back up their understanding of the industry, their opinions often don't get the credit they deserve. Grass-roots organizations like the Maine Lobstermen's Association and the Down East Lobstermen's Association have been instrumental in helping fishermen unite and fight to have their voices heard. The success of the Maine lobster industry over the last thirty years is a just reward for the hard work and conscientiousness of lobstermen and lobster associations along the coast of Maine.

Thankfully, there now seems to be more of a pattern of scientists, government and fishermen working in cooperation to understand what might be happening on the ocean floor. One example of this is the Zone Management Law. Passed in 1995, it has helped improve stewardship by creating local management units (zones) that are ecologically distinct and have elected fisherman representatives.[58]

Yet the Zone Management Law has also brought with it a slew of other, more contentious measures, including trap limits, an apprenticeship program and rules that limit entry into the fishery. Most recently, new trapline regulations have cost fishermen thousands of dollars in replacement rope. At the same time, scientists continue to warn about the impending demise of the industry. Currently, the concern is that the Maine lobster industry is *too* successful. The collapse of other ground fisheries due to mismanagement means that the Gulf of Maine today is an ecosystem dominated by lobsters. At the same time, the Maine economy has become precariously dependent on the lobster industry. Some of the less wise fishermen operate in heavy debt because they, and the banks, believe the golden age won't come to an end. If a biological or economical calamity were to occur, Maine would be hit hard.

There is no denying that the Maine lobster industry is a fragile one. The biggest threat, though, is in the eye of the beholder. For fishermen who prize their independence and feel the tradition of lobstering is their own, the biggest risk is government-imposed regulations that could have a negative impact on the industry. Most of the older or wiser fishermen don't worry about downturns in the lobster stock. A drop in abundance would be challenging, but it would also help with self-regulation, driving out the less dedicated fishermen and leaving those truly committed to carry on with the fishing tradition.

HOW TO BECOME A LOBSTER FISHERMAN

You get your first pair of rubber boots when you're four or five, and your mother's lucky if she can pry them off you to get you to bed.[59]
—*Vinalhaven lobsterman Walter Day*

Lobster fishing delivers tremendous levels of satisfaction for people who prize their independence. Long days of manual labor, uncertainty and risk are a small price to pay for those who take pleasure from self-employment and fresh sea air. When asked to rate their level of job satisfaction, over half of the Down East fishermen in my survey gave themselves the highest possible score. Even those who didn't were still likely to rate their satisfaction as eight or nine out of a possible ten. These men aren't just the masters of their ships; they are the captains of their souls.[60]

Not only does lobster fishing deliver high levels of satisfaction, in the last several decades, it has been able to provide some decent paychecks as well. Each year, a number of enthusiastic tourists contemplate the possibility of giving up corporate America, buying a boat and living the "simple life" of a lobster fisherman. How romantic. Yet lobster fishing is by no means a simple industry to break into. When people ask me what it takes to become a lobster fisherman, I offer the following advice:

1. Be Born into It
The easiest way to become a lobster fisherman is to be born into it. Lobster fishing is traditionally handed down from generation to generation. Of the fishermen I surveyed for this book, almost 70 percent had a father or grandfather who was a lobster fisherman.

Above: My brother fishing with my father in the early 1980s.

Right: My brother teaching his son, Jackson, about lobster fishing.

Being born into an established fishing family delivers a degree of credibility that is impossible to earn any other way. Even newcomers who marry into a fishing family are not welcomed in the same manner as the son or daughter of a reputable fisherman. This son or daughter is seen as a trusted member of the community and deserving of the local resources. He or she effectively "inherits" the right to fish the area.[61]

Those who are brought up in a lobster-fishing family will also inherit a wealth of priceless knowledge about how to fish the local area and conserve the resource. This knowledge will often be complemented by practical experience, having fulfilled sternman duties on their father or grandfather's boats. By the time the son or daughter is ready to become a full-time fisherman, he or she will have had over ten years of education about the trade and a well-founded respect for the industry.

2. If You Weren't Born into a Fishing Family, Integrate Yourself into a Fishing Community

If you want to try your hand at lobster fishing but were not lucky enough to be born into a family of fishermen, the first and most important thing to do is integrate yourself into the community where you wish to fish. Only when you become a full-time resident of a lobster fishing village will you be welcome to fish there. Outsiders who start fishing an area are perceived as robbers of the local resources. Their traps won't last long in the water.

3. Start Young and Start Small

The most respected fishermen are those who enter the industry slowly, fishing a few traps from a skiff or small boat before graduating to a big boat and full gang of traps. This approach allows the fisherman to learn as he goes and earn the respect of his fellow fishermen by demonstrating his commitment to the industry. A modest start also allows a fisherman to establish himself financially. Starting young and on a small scale means the fisherman can build up his business with relatively little debt. Then, when it comes time to advance to a full-scale lobster boat, which can cost over $200,000, he can use the equity from his smaller boat and any savings he's made from those first years of fishing to help finance this investment.

As I've covered in previous chapters, many Maine lobstermen begin their careers young by going sternman with their fathers or grandfathers. One-third of those I surveyed for this book started lobstering before they'd even reached their teens. By the time they are in their teens, they will often begin fishing a small gang of traps out of their own skiff or outboard boat. In their

Right: Josh Cates and his son, Lucas, ready to go lobster fishing. *Courtesy of Laurie Cates.*

Below: Josh Cates helping his son, Lucas, haul traps from a skiff. *Courtesy of Laurie Cates.*

late teens or early twenties, those serious about the industry will graduate to a full-scale operation.

The State of Maine has recently created several laws that in some ways complement and in other ways conflict with the traditional patterns of entry into the lobster-fishing industry. Youngsters can obtain their first lobster-fishing license, called a student license, at the tender age of eight. With that license, the student can work as a sternman and can also fish a small gang of traps of his own. He must haul his own gang of traps from his own boat—a rule that reinforces the owner/operator nature of lobster fishing but can also sometimes prevent a father from helping his son. In addition to fishing from his own boat, the youngster must commit to fishing year after year in order to build up the number of traps in his gang. At the age of eighteen, the fisherman can graduate from a student license to a commercial license and fish a gang of three hundred traps.

Cutler fisherman Neil Corbett (on the right), who fished until the age of eighty-five, with his sternman Olye Huntley.

4. If You Didn't Start Young, Become an Apprentice

Adults who want to get into the lobster-fishing industry but didn't start young and work their way up are now legally required to complete a two-year apprenticeship program. During this time, the apprentice must log one thousand hours of time, over two years, fishing as sternman for another licensed lobsterman. In addition to learning the craft, the apprentice will likely be paid for his efforts. Traditionally, sternmen are paid a percentage of the catch. This percentage ranges from fisherman to fisherman but is often within the 10 to 20 percent range.[62]

Completing an apprenticeship program alone will not guarantee you entry into the lobster-fishing industry. Maine lobster fishing is now a limited-entry fishery, meaning that often one fisherman must retire in order for another to gain admission to the industry. And older fishermen are often in no rush to retire. There is a saying that "Old fishermen never retire, they just get a little dingy," and this saying certainly rings true for Down East lobstermen. Often, the completion of an apprenticeship program is followed by several years on a waiting list before receiving a commercial licence.

MAINE LOBSTER BOATS THROUGH THE AGES

The absolutely lovely characteristic of a lobster boat is the relationship between boat speed and engine rpm. For each application of the throttle, you get a commensurate increase in speed.
—Spencer Lincoln, Down East yacht designer[63]

The lobster boats of today are sleek, fiberglass affairs with fresh colors and modern electronics. However, most of the last century of lobster fishing has played out with a variety of more basic boats. During this time, the biggest design changes to lobster boats have been the introduction of engine power in the early 1900s and the development of fiberglass in the 1970s.

DORIES: THE FIRST LOBSTER-FISHING BOATS

The very first lobster boats were dories—flat-bottomed, lapstrake-sided boats powered by oar or sail. The Banks dory, made famous in paintings by Winslow Homer and others, was one of the true workhorses of early American coastal life. A unique Yankee invention, the dory has been used for fishing since at least the 1700s, and the earliest known moving film footage from Maine, shot in 1901, shows a Maine fisherman hauling square-sided wooden traps aboard a dory.[64]

Lobster fishing during the dory days. *Courtesy of the Penobscot Marine Museum.*

PEAPODS

The next evolution in lobster-fishing boats came with the advent of the peapod. Annual reports from the United States Commission of Fish and Fisheries from the late nineteenth century indicate that the peapod was developed on the island of North Haven in the 1870s. Other Maine islands have been instrumental in its evolution.[65]

The shape of the boat is a reflection of its name. The double-end design allowed the peapod to be easily rowed in both directions. Its shallow draft and maneuverability made it ideal for working around Maine's rugged coastline, where larger, deeper boats could not venture. Most commonly, these boats were propelled by oar, with the fisherman rowing from a forward-facing position, though some models were designed to be powered by sail as well as by oar. Even today the peapod is a popular boat, having found new life as a family pleasure boat.

Cutler Harbor during the days of fishing by oar and sail. *Courtesy of Pauline Cates.*

THE FRIENDSHIP SLOOP

The Friendship sloop was the favored boat design for lobster fishing on the Maine coast in the years preceding the debut of gasoline engines. Powered by sail and easy to maneuver, the Friendship was often worked by just one man when hauling lobster traps. The sloop boat design dates back to the 1850s and came to be known as the Friendship sloop in the late nineteenth century, given that the boat shop of one of the principal designers, Wilbur Morse, was located in the town of Friendship, Maine.[66]

1910 "MAKE AND BREAK" ENGINES

Around the turn of the twentieth century, one- and two-cylinder engines with "make-and-break" ignition systems were introduced and quickly adopted by

Finishing up a day of fishing in the early 1900s. *Courtesy of Pauline Cates.*

lobstermen up and down the Maine coast. These boats made it possible to go fishing without wind and greatly expanded the range a lobsterman could fish as well as the length of the fishing season. These engines inspired better hull designs to take advantage of the engine's potential. Better hull designs inspired the development of more sophisticated engines to maximize hull performance. And so the dance between hull design and engine performance continues to this day.[67]

THE LEGENDARY JONESPORT-BEALS TORPEDO STERN

"If you want a lobster boat, go to Beals Island, cuz you can kick over a stump and find a boat builder there." This was a common claim during the halcyon days of wooden boat building in Maine. Though Beals Island is barely 5.7 square miles, with a population of under one thousand, the island housed a dozen boat shops up until about 1980. Most Beals Island boat builders didn't learn their trade in school. Rather, their skills developed through the years, from their practical knowledge of what made a good working lobster boat and their cultural practice of passing expertise along to the next generation. Beals boat shops were social gathering places for friends and family, particularly sons and sons-in-law. Some children literally grew up on the shop floor absorbing the craft. The boat shop was their school.[68]

While boat building has been a fixture of Jonesport-Beals since the 1700s, in the early 1900s, a couple of events conspired to plant Jonesport-Beals firmly on the boat-building map: the arrival of the "Wizard of Beals" and Prohibition. Around 1912, William "Pappy" Frost moved to Beals Island from Nova Scotia. Frost was a third-generation boat builder and a master at matching hull speed to horsepower without any compromise to the utility required in a lobster boat. He appeared at a critical time in boat-building history, when fishing boats were transitioning from sail power to gasoline engines, and focused on developing hull designs that would function more effectively with these engine-powered boats. Known as the Wizard of Beals, by 1920 Frost's bespoke hull design—with a flatter aft section and "torpedo" stern—was performing better than most other boats powered by the gasoline engine.[69]

By the late 1910s, another transformation was taking place in American society; in 1919, the U.S. Congress passed the Volstead Act, kicking off the era of Prohibition. Ever the opportunists, Mainers turned Prohibition into an economic opportunity. When ships from Canada, loaded with liquor, anchored just beyond the three-mile limit, fishermen, businessmen and

My great-grandfather's and grandfather's wooden lobster boats in the mid-1900s.

A Jonesport-Beals lobster boat. Built on Beals Island by Osmand Beal in 1985, the *Gerry Ann* is the only remaining wooden boat in Bunkers Harbor. The boat belongs to John Chipman Sr. *Courtesy of Theresa Chipman.*

the underemployed went out to ferry the cargo ashore. These rumrunners, needing to stay ahead of the government patrol boats, began placing orders for Frost's fast boats. Frost responded to their demands by continuing to perfect his craft, supplying them with bigger engines and better boats. The government, realizing it could not catch Frost's boats, began buying his designs as well. At one point during Prohibition, Frost was building rumrunners at one end of his boatyard and boats for the Coast Guard at the other end. This cat-and-mouse game was not just great business for Beals boat builders, but it also forced them to raise the game on boat design.[70]

The Wizard of Beals is said to have crafted up to one thousand boats during his life, with many of those boats built on Beals Island. In addition to the torpedo design, he was also responsible for the small, narrow "brimstone" vessel, which is possibly the most beautiful of all lobster boats. His designs were brought to life and advanced throughout the twentieth century by master craftsmen on Beals Island such as Ernest Libby Jr., Calvin Beal and Osmond Beal. While Prohibition ended in 1933, the passion for fast lobster boats continues to this day. Outrunning the Coast Guard has been replaced by friendly lobster boat racing competitions up and down the coast of Maine, where boat builders compete to see which of their designs is fastest.[71]

Fiberglass Boats: Light, Fast and Sophisticated

When my father started lobster fishing in 1976, fiberglass boats were just beginning to replace traditional wooden boats. The first fiberglass boats on the market were met with great skepticism by many lobstermen. Referred to by some fishermen as "Clorox jugs," many worried the boats would pound you to death in a sea. These criticisms were not unfounded, mostly because the first fiberglass hulls weren't nearly as well crafted as those of wooden boats.

That said, even the first, poorly designed fiberglass boats had a major advantage over wooden boats. They required considerably less maintenance. Even a well-built wooden lobster boat leaks a bit. For this reason, many fishermen would take their boats out of the water during the winter. Each spring, the outside of the boat would need to be repainted from top to bottom, and the inside of the boat would need to be repainted every three to five years. The allure of less maintanence led fishermen to start gravitating toward fiberglass boats in the early 1970s and asking for more speed and seaworthiness out of the hull design.

My brother and me on the bow of my father's first Young Brothers boat, the *Celia Marie*. Surrounding us are trophies won by racing the boat at the Maine lobster boat races.

Some of the first fishermen to start asking more out of fiberglass lobster boats were the Young Brothers of Corea, Maine. In 1973, Arvid Young tried out his first fiberglass boat, saw how poorly it handled versus the finely crafted wooden hulls he was used to and spotted an opportunity. Soon after, he and his two brothers took a mold off a Beals wooden design and pulled their first fiberglass hull. The day they launched this first boat, six others were sold. Shortly after, they began working with the talented Ernest Libby Jr. of Beals, who perfected the design of their thirty- to forty-five-foot hulls. Libby was incredibly skilled at being able to put a bottom on a boat that was fast while still being steady and seaworthy. With Libby's help, the Young Brothers' reputation for producing swift, stable fiberglass boats grew quickly. This reputation was fueled in part by showing off their designs up and down the coast at Maine lobster boat races. Fishermen bore witness to their race-winning boats (one achieving a speed of sixty-four miles per hour) and thought, "I'll have me one of those." The rest is legend, as the Young Brothers built over 550 boats at their Corea yard. My father ordered his first Young Brothers hull in 1978. He fished that boat for five years, winning eight lobster boat races with it.[72]

Modern Marine Masterpieces

As the more sturdy, and soon faster, fiberglass lobster boat came into its own along the coast of Maine, engine choices began shifting from automobile engines to proper marine engines, and the entire electronic and hydraulic system reached a new height of sophistication. Before the fiberglass boat boom, most lobster boats weren't properly rigged for marine use. Most fishermen would use a retrofitted car engine that had passed its prime on the road. My grandfather even sourced some of his boat engines from our town dump. When he was in need of a "new" engine, he'd go out to the Cutler dump, roll over an abandoned car and nab the engine.

The move from gasoline car engines to modern marine engines meant longer-lasting motors and greater fuel efficiency. Where a gas engine would often have to be replaced every season, a good diesel engine will last a fisherman up to ten years. The year my father transitioned from a 455 Oldsmobile gas engine to a comparably sized diesel engine he was also able to cut his fuel consumption by about two-thirds.

As engines evolved, so did the electronics. Manual bilge pumps were replaced with electronic models. Watertight floors paved the way for circulating saltwater systems, which helped keep the lobster catch fresh and the floors and decks clear of bait juice and sea slime. Winch heads were replaced with sophisticated hydraulic pot haulers.

Up until the late 1960s or early '70s, if you were to step on a Maine lobster boat, you'd still likely find a small mast near the stern. On windy days, a sail would be hoisted to reduce the effects of wave motion, thus providing a more stable working platform. You also might find an old ceramic washtub in the wheelhouse filled with lobster bait. These old washtubs often found a second life on lobster boats, as they didn't rust. You might even find two steering stations—one alongside the hauler, referred to as the break arm, which would be used for steering while hauling traps, and another, a proper steering wheel, down forward for cruising.

STANDING AT THE HELM OF A MODERN-DAY LOBSTER BOAT

Step aboard a lobster boat today and you'll be greeted by sleek, fiberglassed paraphernalia and sophisticated electronics and hydraulics. Below decks sits an engine that costs more than a brand-new pickup truck. Next to the steering wheel is a hydraulic hauler so powerful it can pull up to two thousand pounds if needed. Mounted to the ceiling of the wheelhouse is a medley of technical equipment for navigating and communicating.

A modern-day lobster fisherman will use at least five pieces of electronic equipment for his fishing. A depth sounder and fathometer will give him a glimpse of the ocean floor, allowing him to distinguish muddy bottom from rocky shoals when placing his traps. A radar unit, loran navigational

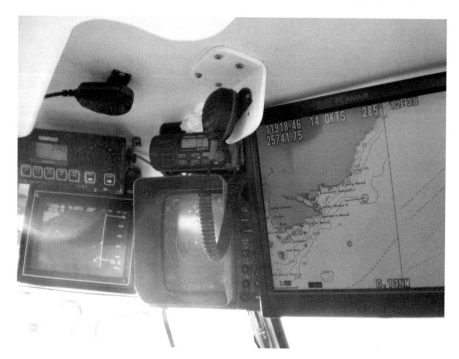

Modern marine electronics.

locator and Global Positioning System (GPS) satellite plotter will help him navigate through dense fog and find the exact coordinates of traps offshore, where there are no landmarks. A VHF marine radio will enable him to communicate with other fishermen out at sea and with loved ones back at home. The simple VHF is actually one of the most vital pieces of equipment on the fisherman's boat, for it allows him to quickly radio to surrounding fishermen for help in the event of a breakdown or injury.

The blare of the VHF combined with the loud rumble of the engine and the piercing whine of the pot hauler make for a very noisy environement while out hauling. Many fishermen add an FM radio to this medley of noises, tuned as best as possible to a local country music station and cranked up to project over the other racket. I have suffered permanent hearing damage from my years working on a boat, as, I'm sure, have many other fishermen.

Farther aft, out behind the wheelhouse, the rigging varies quite a bit from boat to boat. Those who fish trawls offshore often kit their boat out with a trawl table or open up the stern, creating an easy way for a trawl to exit the boat. Some fishermen also rig up a dip tank, a container of hot salt water, to keep their buoys clean. Buoys, like anything that sits for a while in the nutrient-rich waters of Maine, quickly become covered by a film of dark green algae, which masks their bright colors. One of the most effective ways to kill off the marine growth is to dunk the buoy in scalding hot water. If installed, a dip tank also tends to serve other purposes on a boat. Some fishermen will use their dip tank to heat up a tin of soup for lunch. I used to plunge my rubber-gloved hands into the dip tank on cold fall days to warm them up.

MODERN-DAY LOBSTER BOAT HULL DESIGN

The hull of Down East lobster boats comes in one of two designs: built-down or skeg. Built-down style boats, usually found south of Ellsworth, are those where the hull esses or curves into the keel. The skeg models are those where the hull goes flat into the keel. Which design is best? There are very vocal camps on either side of the question. I grew up in skeg country, so my family is a bit biased toward that style of boat. Built-down boats claim to be more buoyant and spacious. The V-like space where the hull curves to join the keel creates buoyancy and is a handy place in which to mount the engine (alleviating the need for an engine box). Skeg boats are less expensive and lighter. They are also said to be faster, as there is less holding the hull back. Some believe a built-down boat is more seaworthy due to the buoyancy. Others think a built-down boat gives you buoyancy where you don't want it, causing the boat to rock and roll a bit. In Down East Maine, the Beals, the Young Brothers and Ernest J. Libby all build skeg boats. Spencer Lincoln's designs are built-down boats.[73]

Beyond built-down and skeg variations, there is really no dominant hull style in Down East Maine. The majority of modern Maine lobstermen prefer a semi-displacement or cruising-style hull because they want more speed. While a boat's speed is determined in part by the engine horsepower, shaft RPMs and size of the propeller, some hull designs cause a boat to top out at a certain speed. With a cruising-style hull, each application of the throttle is met with a proportional increase in speed, allowing many boats to achieve a cruising speed of sixteen to twenty-two knots, with top speed on some boats exceeding forty knots.

The last twenty years or so have also seen hull designs trend toward a wider stern to accommodate the larger gangs of traps being fished. To minimize the impact on the boat's seaworthiness, modern-day boat builders have revised their designs, adding deeper keels and larger rudders to help the boat track well in a following sea. Still, every boat is a bit of a compromise. Today's wider-sterned boats allow you to bring five or six tons of gear aboard when you're setting off, but they don't handle as gracefully in a following sea or rough weather.[74]

IT'S NOT ALL HARD WORK

The days are long and the trap hauling routine can often feel relentless. Yet the life of a lobsterman isn't soley hard work. Socializing is a vital part of life for Down East fishermen. In towns like Cutler, where there is no general store, coffee shop, pub or other proper public gathering place, informal spaces spring up where men congregate and gossip. In Cutler, we have "the bilge," the basement of a local fisherman's house, where lobstermen gather for coffee and donuts. Beyond the suggestion of leaving a donation, the refreshments are free. The oldest fisherman to visit "the bilge" even has a special assigned seat—an old barber's chair perched in the middle of the room.

Beyond daily socializing, Down Easters have some special pastimes, born of our lobster-fishing lives. I have included the ones that I think are unique and noteworthy.

THE MAINE LOBSTER BOAT RACES

The whole course is only 'bout an honest three-quarters of a mile. Basically it's nothing but a big drag race. You just kinda point it and punch it.
—*Glenn Holland, master boat builder and owner of the* Red Baron[75]

Starting in late June, harbors up and down the coast of Maine are temporarily transformed into drag strips, and spectators line the shores to watch fishermen battle it out for first place at the Maine lobster boat races.

The competition is quite heated, with some of the souped-up lobster boats reaching speeds of over sixty miles per hour. The races often begin with small, outboard-powered skiffs and then graduate to larger gas- and diesel-powered vessels. As the day progresses, so does the race intensity. Bigger boats with higher levels of horsepower roar down the harbor, leaving giant rooster tails in their wakes. The last race is usually a full-throttle free-for-all for the coveted title of fastest lobster boat.

Lobstermen have always enjoyed a bit of friendly competition. Even in the schooner days, fishermen would often race one another home after a day out at sea. In the 1920s, these races started to become more organized affairs, with Jonesport and Beals leading the way. Between these two fishing villages lies a sheltered body of water called Moosabec Reach. When moter power replaced sails in the 1900s, the Reach became an ideal testing ground for boat builders to trial their latest designs. Each time a builder won the coveted prize of fastest lobster boat, it was a wonderful advertisement for his business. Each time he lost, it motivated him to improve his design. Win or lose, the races were a great way for fishermen and boat builders to blow off some steam.[76]

Today, many fishermen and spectators follow the Maine race "circuit," which stretches up and down the entire coast, running from late June through the end of August. No one gets rich from the sport. Though top prizes can range from $500 cash payouts to new electronic equipment, more common winnings include cases of motor oil, convenience store gift certificates and bushels of foul-smelling bait. Yet most fishermen would agree that lobster boat racing is not about the prizes; it's about the bragging rights.

Bragging rights and pride have been more than enough motivation to fuel notorious rivalries and extreme racing behavior throughout the history of the sport. In earlier years, it wasn't uncommon for a fisherman to smash out his cabin windows and throw fishing supplies overboard mid-race to increase his speed. If the race was close, some fishermen would start tossing over anything that wasn't nailed down to increase their advantage. In fact, at one point, race officials put a rule in place requiring that race boats must have all cabin windows intact in order to compete. The shame of losing a race sometimes compelled a fisherman to sell his boat on the spot. A family friend of ours bought such a lobster boat in the 1960s. The boat had just lost a Jonesport-Beals race, and the owner wanted to offload it. The boat had no windows or floorboards because the owner had smashed and thrown them out during the race.

During the 1980s, there was a particularly heated rivalry between several boat builders along the coast of Maine. One of these boat builders, Glenn

Holland, built and began racing a boat called the *Red Baron*. His vessel reached top speeds of over sixty miles an hour and gained such a reputation that Maine toy makers continue to sell wooden models of it today. Another boat-building family, the Young Brothers, challenged Glenn Holland with their own super-powered lobster boat. They named their boat the *Sopwith Camel* after the famed World War I British biplane—which shot down Manfred von Richthofen and his "Red Baron" German aircraft. The *Camel* could also achieve speeds of sixty-plus at full throttle and had a special seatbelt affixed to the bulkhead to keep the driver secure during a heated race. While both the *Red Baron* and the *Camel* were working lobster boats, they, like most other boats, went through a lot of tinkering at race time. This tinkering could range from changing out propellers and rudders to pumping nitrous oxide into the fuel to increase engine horsepower or reducing weight by removing the entire engine cooling system and simply running raw salt water through the engine.

Though the *Red Baron* and the *Camel* are now retired, the racing circuit is as lively and competitive as it ever was. In Down East Maine, the Jonesport, Stonington and Winter Harbor races tend to draw the biggest crowds, with a festive, sometimes rowdy atmosphere, many a spectacle and the occasional mishap. One year in Searsport, a seaplane raced against the lobster boats. Another year, a boat flipped over mid-race due to the chop. A

The *Camel* under full steam at the Maine lobster boat races.

steering malfunction caused a boat to drive full throttle into the Jonesport-Beals bridge abutment one July, flinging the crew overboard in the process. Thankfully, all survived.

A more common sight at the Maine lobster boat races is a crowd of fishermen and race fans partying it up in the harbor. While some lobstermen bring their boats to the races to rev up their engines, others simply come to "raft up" (tie their boats together) and host a group of spectators. Grills, picnic tables and coolers of Budweiser are brought aboard, and it can turn into one heck of a party.

LOBSTER BOAT PICNICS

One nonwork tradition, in my harbor at least, is to pile a bunch of friends, family and food on the boat on a warm summer afternoon and head to a nearby island for an afternoon beach picnic. Is this the lobster fisherman's equivalent of a bus driver's holiday? In some ways, yes. It requires firing up the engine and using the heister to transport goods onto the boat, but it also gives the fishermen the chance to enjoy and perhaps even show off their boats to others.

For those from away, the words "coastal picnic" might conjure up images of a New England clambake complete with fine wine and microbrew beers. For the average Down Easter, it means hot dogs roasted over a roaring beach fire accompanied by macaroni salad, chips and a variety of home-cooked deserts, courtesy of the fishermen's wives. Entertainment includes rock-skipping contests, local storytelling and possibly even a hot dog–eating contest. On some Cutler boat picnics, there have been over one hundred people in attendance—almost one-fifth the entire population of the town. Ages have ranged from one year old to people in their eighties. There are always at least two dogs in attendance. A beach picnic is one of my favorite ways to spend a sunny summer afternoon in Maine.

LOBSTER CRATE RACES

In addition to boat picnics, Down East fishing families look forward to and take great pride in their summertime Fourth of July and Harbor Day celebrations. Jonesport-Beals and Cutler in particular are famed for their unique and festive celebrations on the Fourth. The pomp and circumstance usually plays out over several days and includes a slew of unique games and competitions that you'd only find in Down East Maine.

In addition to the typical parade, flag raising and singing of "The Star-Spangled Banner," fishermen often dress up their boats with patriotic decorations and sometimes organize a Blessing of the Fleet. But perhaps the most unique event of all is the lobster crate race. As mentioned previously, fishermen used to store their lobsters in five-cubic-foot wooden crates. These wooden crates, when empty, will float on the surface of the water—just barely. During Fourth of July celebrations, towns like Cutler and Jonesport will string a bunch of these crates together in a line, tether them between two wharves and then challenge spectators to run across without falling in the water.

Contestants must be incredibly swift and sure-footed in order to make it across the bobbing string of crates. If a crate runner moves too slowly, the

Crate racing on the Fourth of July in Cutler, Maine. *Courtesy of Janine Drouin.*

Lobster boats decorated for the Fourth of July and lined up for the annual Blessing of the Fleet.

crates will sink beneath his or her feet. If the runner's foot doesn't land in the center of a crate with each stride, that crate will tip sideways and fling him or her into the icy ocean water. Given the odds of making it across the crates are less than 10 percent, you'd think people would be reluctant to sign up for the challenge. Yet the opposite is true. In Cutler, we actually have to cap the number of contestants at fifty each year. Some of these crate runners are visitors from away, while others are Cutler natives. A few of the runners have a ten-year track record of making it across the crates. With water temperatures of fifty-four degrees, a crowd of over one hundred watching from the shore and twenty dollars waiting at the finish line, the stakes are high. One Fourth of July, the stakes were raised even higher by a man who promised his girlfriend he'd marry her if she could make it across the crates. The man was a notorious bachelor, and his girlfriend was eager for him to commit. Desperate though she was, she was not able to make it across the crates. The two never married.

Lobster Trap Christmas Trees

During the holiday season, a large, brightly lit Christmas tree is a common sight in city and town squares throughout the United States. Several towns on the Maine coast have put a unique spin on the tradition by constructing their trees from piles of lobster traps. In recent years, some of the top lobster-fishing ports in New England have even been claiming bragging rights about which has the biggest and best lobster trap tree.

The trap-to-tree tradition started in Gloucester, Massachusetts. Now both Rockland and Beals Island erect trap trees at Christmas time, with each town claiming its tree is superior. Beals boasts that it has the tallest lobster trap tree "in the world." In 2012, it reached sixty feet in height. Rockland claims its tree is the best due to the artful construction, built using a "secret" engineering formula that means it is free-standing—no need to be tethered to the ground.[77]

Behold, the "World's Largest" lobster trap tree, Beals Island, 2011.

A mini lobster trap tree built by my nephews, Ryan and Jackson Lemieux. *Courtesy of Belinda Lemieux.*

The trees are typically decorated with lights and brightly colored lobster buoys. Tree toppers have ranged from a five-foot fiberglass lobster and buoys arranged in the form of a cross to a mannequin dressed as a fisherwoman wearing yellow oilskins, looking out to sea as if waiting for her fisherman husband to return home.

Festive celebrations often accompany the ceremonial trap tree lighting. Beals holds an annual Christmas flotilla consisting of decorated boats that traverse the Reach to the accompaniment of carols and fireworks. Smaller versions of lobster trap trees often adorn the front lawns of fishermen's homes.

HOW TO EAT A LOBSTER

BUYING FRESH LOBSTER

Picking out a live lobster from a fishmonger or off a fishing wharf can be a daunting experience. What's the best size? Should you go for shedder or hard shell? How do you know it's healthy? Here are some things to look for or consider.

Color

If you're only used to seeing cooked lobster, you might be surprised by the dark green/black color of a live lobster. Lobsters only turn red once cooked. The reason for this color change is that, during cooking, all the pigments that make up a lobster's shell are destroyed by heat except the red pigment.[78] The color of the lobster does not affect the taste or quality of the meat, so don't be put off by any strange shades. While most lobsters are the dark green/black color I've described, other lobsters have a more brownish, bluish or yellowish color.

Lobster Size

While many people love the idea of getting their hands on a nice, big lobster, smaller lobsters are actually sweeter in flavor. The smallest lobster you can buy in Maine will weigh around one and one-fourth pounds. I recommend purchasing lobsters in the one-and-one-fourth- to two-pound range to get the best-tasting meat.

How to Eat a Lobster

Hard Shell or Shedder

As above, if you want a nice, sweet lobster, I'd go for a shedder. You obviously have to be in the right place (Maine or New England) at the right time (late summer or early fall) to get a shedder, but if you can get your hands on one, you can't beat it for taste.

If you want the sweet taste of a shedder but a bit more meat, then look for a hardened-up shedder. A few months after it's shed, the lobster's shell will harden up so much that it's quite difficult to determine whether or not it's a shedder by the traditional squeeze test. At this point, the lobster has started filling out the shell a bit, so there's a bit more meat inside, though this meat is still very sweet. You can ask the fishmonger or dealer for a good, hardened-up shedder or look for one yourself. Some fishmongers may be inclined to pass off a hardened-up shedder as a hard shell because they can get more money per pound for the lobster and an unseasoned tourist won't know the difference. To determine whether it's a hardened-up shedder or a hard shell, flip the lobster over and look at the underside of the claws. The claws of a true hard-shell lobster will be a dark, blackish color and scarred from months of fighting off predators and cruising around on the ocean floor. The underside of a shedder's claws is bright orange in color and free from the dark scars you see on a hard shell.

Male or Female

While many people don't fuss over the sex of the lobster while picking it out, there are some nuances to consider. Female lobsters have larger tails and, therefore, more meat. At the same time, some people (myself included) feel that a female lobster loses some of its taste when pregnant. While a pregnant lobster can't be sold on the market once it's "egged out" (once the eggs are showing on the tail), the lobster could be pregnant for several months before these eggs appear.

Health/Vitality

Make sure that you choose a lobster that is lively and active. Most lobsters will flap around a bit when they're picked up. Even if they don't flap around, the tail should be either straight or curled under when picked up. If you want to make sure the lobster is healthy, check its tail. If you uncurl the tail of a healthy lobster, it should spring back to a curled position. If you hold the lobster up by the body (carapace) and turn it over so that its legs are facing upward, the tail should remain tense and extend outward or, again, curled up. If the tail flops down loosely, the lobster is likely dead or not far from it.

Never purchase a lobster you suspect is dead. If a lobster dies once you purchase it, it's generally safe to cook and eat within twenty-four hours. If you're not sure when the lobster died, don't risk it.

Another warning sign that a lobster is not healthy is the integrity of the tail meat once cooked. If the tail meat comes out of the lobster in one chunk, it's healthy. If the tail meat is crumbly, the lobster has spoiled and is not fit to eat.

HANDLING LIVE LOBSTER

Live lobsters are very perishable and require a controlled saltwater environment to remain alive. They do not generally live much more than a day out of water, so ideally, you should cook your lobsters on the day you purchase them.

The best way to keep lobsters alive at home is to refrigerate them. My family simply puts our lobsters in a crisper drawer until we're ready to cook them. Other people cover the lobsters with moist seaweed or a damp cloth and leave them in the refrigerator until ready to cook. Never place lobsters in tap water to try to keep them alive; lobsters are saltwater creatures, and fresh water will kill them.[79]

KILLING THE LOBSTER

Killing lobster is not a task I relish, so I try to get the deed over with as quickly as possible. Sometimes I flip the lobster over and stab it right below the mouth with a knife. This method kills the lobster instantly. Other times, I incorporate the killing into the cooking process by holding the lobster nose down in the steaming water. The lobster inhales the steam and perishes within a few seconds.

If you go with the steam inhalation method, be sure to put an oven mitt on your hand to protect your hand from the steam. Also, if you're cooking a bunch of lobsters in one pot, you'll need to kill each lobster before filling up the pot. Otherwise the lobsters at the top will not be subjected to enough steam to kill them straight away, and they will likely flap their tails around for several minutes while they are dying.

The Lobster Institute research organization claims this tail flapping is simply a reflex action found in lobsters, known as the escape response. Still, it can be rather distressing to hear. The institute actually recommends chilling the lobster before dropping it in water that has come to a rolling boil as the best way to minimize movement and noise while cooking.[80]

With regard to noise, one question many people ask is whether lobsters scream when they're put in a pot of boiling water. The answer is no. Lobsters actually don't have vocal chords or any other means of vocalization. According to Dr. Robert Bayer, a professor of animal and veterinary sciences at the University of Maine and director of the research organization the Lobster Institute, any noise you might hear while a lobster is cooking is likely air coming out of its stomach through its mouthparts.

It's also important to mention, as confirmed by the Lobster Institute, that lobsters do not feel pain when you cook them, as they have no brain and a very simple nervous system. In order for an organism to perceive pain, it must have a more complex nervous system.[81]

COOKING LOBSTER

Lobster can be boiled, steamed, grilled or broiled. I prefer steaming lobster, as it is a more gentle method of cooking lobster, which results in slightly more tender, flavorful meat. Not only does steaming preserve a little more of the flavor, but this method is also more forgiving on the timing front, as it's harder to overcook a steamed lobster. Below, I've provided my approach to steaming lobster, along with directions from the Lobster Institute on boiling and grilling lobster.

Steaming

To properly steam a lobster, fill a large pot with about two inches of water, cover the pot with a lid and bring the water to a steam. While many people will recommend steaming lobster in sea water (claiming it helps maintain its characteristic ocean taste), my family has always steamed our lobster in fresh water.

Place the lobster in the pot, keeping the bands on. Replace the lid and cook for twenty minutes for a one-and-a-half-pound hard-shell lobster. If the lobster has a soft shell, you can reduce the cooking time by several minutes. You'll know the lobster is cooked when it turns bright red in color.

Boiling

Fill a large pot approximately three-quarters full of seawater or salted water (two tablespoons of salt per quart of water). Use about two and a half quarts of water for each lobster. Bring the water to a boil. Put in the live lobsters, one at a time, cover and bring the water to a boil again. Then, lower the heat and simmer about fifteen minutes for a one- to one-and-a-quarter-pound hard-shell lobster and twenty minutes for a one-and-a-half-pound hard-shell lobster. For soft-shell lobsters, reduce the cooking time by three minutes.[82]

Grilling

Parboil lobsters in boiling water for five minutes. Remove the lobsters and immediately put into a large pot/bowl of cold water to stop the cooking. (You can drain the lobsters and store in the refrigerator if you do not plan to grill them right away.)

Using a sharp knife, slice the lobster down the middle (easiest to cut legs side up). Remove the black vein from the tail, the greenish tomalley from the body and the sand sac located near the head. Baste the lobster meat with some oil or melted butter.

Grill the lobsters flesh side down for five to six minutes, or until the flesh is just beginning to look opaque. Turn the lobsters over, baste with more oil and continue to cook for four to five minutes longer, or until the lobsters are cooked through.

PICKING OUT AND EATING LOBSTER

The first thing to say about picking out lobster is that everyone has his or her own method. This is how I was taught to eat lobster from the shell and how I continue to teach others.

The second thing to note is that a boiled or steamed lobster retains water when cooked, and as you break the lobster apart, it will "leak" this hot water. Allow ten minutes or so for the lobster to cool before cracking it open, and be sure to do so while holding the lobster over your plate (or even over the kitchen sink if you're eating the lobster at home) to avoid extra mess.

The final thing to mention is that some people pick out all the meat and discard the shell before eating while others prefer to eat as they pick. Eating lobster from the shell is not an exercise in etiquette, so choose whichever option works best for you. I prefer to eat as I pick, but that's probably because I'm just too impatient!

Equipment needed
- Lobster cracker (while you can get special "lobster crackers," nut crackers work just as well)
- Lobster pick (a slender instrument for extracting meat from the small joints)
- Bucket or bowl (for discarding lobster shells)
- Lobster bib (optional, but if it's your first time eating lobster and/or you're in a restaurant, a bib is a good idea)
- Paper towels (lots of them!)
- Melted butter or lemon (as a dressing for the lobster)

Instructions for Picking Out a Lobster

- Hold the lobster by the back (carapace) with one hand. With the other hand, grab the legs (there are four on each side) and tear them off with a gentle twisting motion. These legs contain lovely little morsels of sweet meat, which you can suck out of the cavity (like sucking on a straw). If you prefer, you can pick the meat out with your lobster pick and then eat.
- Continue to hold the lobster by the back with one hand and, with the other, tear each of the claws off by gently twisting the claws at the point where the claw knuckles attach to the body.
- Extract the meat from the claw knuckles by using the cracker to snap apart each of the knuckles. Then use the lobster pick to scoop out the meat from each knuckle. This is some of the sweetest meat of the lobster. Enjoy.
- Next, use the cracker to break open each claw. Hold the tip of the claw in one hand and, with the other hand, place the cracker sideways over the fattest part of the claw and squeeze hard (especially for a bigger, hard-shell lobster). Once you've made a big crack in the claw, you can usually snap it apart with two hands and easily scoop out the meat. The meat from one of the claw tips should slide out, along with the bulk of the claw meat. To get the meat out from the other claw tip, snap it off from the rest of the claw.
- To remove the lobster tail, grasp the tail in one hand and the back of the lobster in the other and then twist in opposite directions. The

tail should easily separate from the body with this motion. Next, break the flippers off the end of the tail with your hands. Once the flippers are removed, you can use a fork or your fingers to push the meat out of the tail (push from the narrow, flipper end of the tail).

- Once the tail meat is removed, you'll want to peel the top, tab-like section of meat away from the tail. Underneath this tab you will most likely see a long, thin brownish strip nestled into the crevice of the tail. You'll want to wipe or wash away this substance, as it is the lobster's waste. If you're picking out a female lobster, you may also find a waxy reddish substance in this section of the tail: the lobster's roe. The roe is edible, though most people choose not to eat it.

- Some diehard lobster lovers (myself included) even tear off the top of the shell (the carapace), split the body lengthwise and pick out the little bits of tender meat where the legs were attached to the body. Warning: this is a very messy process.

- If you venture inside the body, you'll notice a greenish, grayish substance, which is the lobster's liver. The liver, called "tomalley," is edible though it has a briny flavor (versus that of the sweet lobster meat), and many people choose to discard it.

- Additionally, as you pick out your lobster, you will notice a white substance around the meat (on the claws in particular). This is the blood of the lobster. It appears clear when the lobster is alive but congeals and turns white when cooked. It is safe to eat though flavorless and easy to scrape off if you wish. Sometimes you will see some of this white blood in your pot as you boil your lobsters. Again, this is quite normal and perfectly harmless.[83]

- If you've been able to resist eating the lobster as you pick, you may want to dip your accrued lobster meat in melted butter or give it a squirt of lemon before eating. I tend to eat my lobster plain so I can enjoy its sweet, natural flavor.

Reheating Cooked Lobster

Once cooked, lobster can be left in the shell for about twenty-four hours. During this time, you can simply reheat it in the shell in the microwave or in steaming water. If you plan to wait more than twenty-four hours before enjoying the rest of the lobster, pick the meat out of the shell and store it in a bowl or plastic bag in the refrigerator. You can then eat it cold—in a salad or lobster roll—or warm by heating the meat in a saucepan with a small amount of melted butter.

How to Freeze Lobster

The following information has been prepared by the University of Maine Department of Food Science and Human Nutrition and the Lobster Institute.[84]

How to Prepare and Freeze Whole "In the Shell" Lobster

Properly prepared whole or "in the shell" lobster has a good quality shelf life of nine to twelve months.
1. Lobsters should be chilled and alive.
2. Blanch at 212 degrees Fahrenheit for sixty seconds in a 2 percent salt brine (2.5 tablespoons of noniodized or sea salt to two quarts of water).
3. Chill after blanching in cold running water or in a tub with a mixture of 50 percent water to 50 percent ice.
4. Following a fifteen- to twenty-minute chill, remove excess surface water.
5. Place in commercial freezer bags and remove as much air as possible. (New Ziploc vacuum bag systems available at supermarkets work well.)
6. Place in a second freezer bag or over-wrap with a laminated freezer wrap.
7. Freeze at -18 degrees Celsius (0 degrees Fahrenheit)—standard for home refrigerator/freezer units.
8. Store frozen at -18 degrees Celsius (0 degrees Fahrenheit) or lower—the lower the storage temperature the better the lobster meat quality will be maintained.
9. Thawing directions: Lobsters should be thawed overnight in the refrigerator.
10. Thawed lobsters should be boiled in a 2 percent salt brine for twelve to fifteen minutes.

You can also just freeze tails and claws "in the shell":
1. Follow steps 1 through 4 above.
2. Remove claws and tail from blanched lobsters.
3. Continue with steps 5 through 10 above.

How to Freeze Picked Tail and Claw Meat

My family has always cooked and picked our lobster meat before freezing it. To do so, we fully cook the lobster, pick out the meat and put it in either a Ziploc freezer bag or, ideally, a vacuum-sealed bag. Before sealing, we add three or four teaspoons of milk. We always consume the meat within several months of freezing it. We find that a bit of the flavor is lost in the freezing process, so we usually turn the meat into a lobster Newburg for serving.

MY FAVORITE LOBSTER RECIPES

THE LOBSTER ROLL CLASSIC

Ingredients
The classic Maine lobster roll can be made with as little as four elements: butter, bread roll, mayo and, of course, lobster.

Directions
Simply cook and pick out the lobster, cut the meat into big chunks, coat it with a reserved amount of mayonnaise and place it in a toasted bread roll. I recommend using J.J. Nissen rolls from Maine, which have exposed sides, perfect for slathering and toasting with butter. If you want to add extras, the most popular are a bit of chopped celery and a dusting of paprika. Serve with some chips and a crisp dill pickle.

THE LOBSTER ROLL FANCY

This recipe is compliments of my friend and chef Natalie Gingerich.

Ingredients
2 tbs. celery chopped fine
1 tbs. parsley minced
juice of half a lemon
2 tbs. shallots minced

2 tbs. chives chopped
1/4 cup homemade mayonnaise
few dashes of hot sauce if so desired
1 cup lobster meat, chopped but coarse
salt and pepper
butter
4 New England–style hot dog buns

Directions
In a mixing bowl, incorporate the celery, parsley, lemon juice, shallots, chives, homemade mayonnaise, hot sauce and lobster meat and season with salt and pepper to taste. Toast bun on the sides with butter until golden brown and fill with about one-quarter of the filling. The perfect complement to this dish would be a nice glass of champagne or a white Belgian beer.

LOBSTER NEWBURG

Lobster Newburg is my favorite way to enjoy lobster. I use my grandmother and mother's recipe, which is as simple as it is delicious. I have served lobster Newburg on many special occasions, including at my wedding reception.

Ingredients (for 2 people)
For the Lobster
2 lobsters (around 1 1/2 pounds per lobster)
2 tbs. butter

For the White Sauce
2 tbs. butter
2 tbs. flour
1 cup milk
nutmeg to taste
red (cayenne) pepper to taste
salt and pepper to taste

Directions
Use approximately one lobster per guest (around one and a half pounds per lobster). Cook the lobster and pick out the meat. Cut the lobster meat into bite-sized chunks and then place in a frying pan on medium heat with

several large chunks of butter. Fry the lobster for about five minutes and then take it off the heat but don't drain.

In a saucepan, on medium heat, make your basic white sauce. To do this, place 2 tablespoons of butter in the saucepan. Once melted, add 2 tablespoons of flour and mix. Once mixed, add 1 cup of milk and whisk the ingredients until thickened. Add a dash of salt and pepper plus a dash of nutmeg and a dash of red pepper. Take the white sauce off the heat and then add in the lobster (including juice). Note, if you're cooking lobster Newburg for more than four people, you'll want to increase the amount of sauce.

Once you have combined the lobster and sauce in the saucepan, cook the lobster Newburg mixture in a double boiler on a very low heat for three to four hours so the sauce has a chance to really absorb the flavor of the lobster. The sauce will turn pinkish as it becomes increasingly flavorful.

When it comes time to serve, I usually toast saltine crackers in the oven until crisp and then pour some of the lobster Newburg over the crackers. While simple, this is how we've served lobster Newburg in my family for generations, and it tastes lovely. Spagetti or linguine are great alternatives to crackers if you so prefer. I usually accompany the dish with peas.

LOBSTER MACARONI AND CHEESE

Lobster macaroni and cheese may sound a bit odd, but trust me, it tastes delicious. It is also a relatively simple way to serve lobster. If you only have a single lobster or a small bit of leftover lobster meat and want to make a meal out of it, this is a great recipe to use.

Also, while you can use your own preferred macaroni and cheese recipe, I prefer the below recipe, as I think gruyere, fontina, cheddar and Parmigiano-Reggiano do a wonderful job of bringing out the flavor of the lobster.

Ingredients
1 to 2 small lobsters or 3 lobster tails
8 oz. macaroni
5 slices crusty white bread for the bread crumb crust
4 tbs. unsalted butter
1/4 cup finely diced white onion
1/4 cup all-purpose flour
3 cups whole milk
1/2 cup grated fontina cheese

1 cup grated Gruyere cheese (reserve ⅓ of this for the topping)
2 cups grated white cheddar cheese, ideally extra sharp (reserve ⅓ of this for
   the topping)
1 cup grated parmesan cheese (reserve ½ of this for the topping)
pinch of cayenne pepper
pinch of nutmeg
salt and pepper to taste

Directions
Steam the lobsters (15 minutes), cool, pick out the meat and then sauté in butter (7 minutes) and set aside.

Cook and drain the pasta. While doing so, preheat the oven to 375 degrees and butter a 1½-quart baking dish.

Make the breadcrumbs by tearing the bread into large pieces (I leave the crust on, as I like the extra crunch), heating up 2 tablespoons of butter and mixing the bread chunks with the melted butter.

Make the cheese sauce by melting the butter in a pot, adding and cooking the onions until translucent and then adding the flour (stirring for about 40 seconds) and then the milk. Once the milk is added, whisk to combine while bringing the mixture to a simmer until thickened (about 5 minutes). Then add the fontina, Gruyere, cheddar and parmesan cheeses (remembering to keep aside a bit for the topping). Stir until the cheeses are completely melted and then season with cayenne, nutmeg, salt and pepper.

Add pasta and lobster to sauce and then pour it all into the prepared baking dish and sprinkle it with the reserved cheeses and then the bread crumbs.

Bake for about 20 minutes until the cheese is golden brown.

LOBSTER RAVIOLI

If you want to impress guests, lobster ravioli is a great dish. The simple combination of lobster, ricotta and chives, along with the champagne sauce, is just delicious. This recipe is compliments of Natalie Gingerich.

Ingredients
2 tbs. chopped shallots
2 tbs. butter
1 cup mascarpone or ricotta cheese
1 cup cooked and chopped lobster meat

2 tbs. chopped chives (any fresh herbs would work well)
¼ cup finely grated parmesan cheese
salt and pepper to taste
1 egg, beaten
1 package wonton wrappers

Directions
Sauté the shallots in the butter and then add to a mixing bowl and allow to cool. Once cooled, add the mascarpone (or ricotta), lobster, chives and parmesan cheese and season with salt and pepper. Taste and adjust seasoning if necessary. Clear some counter space and then lay out the wonton wrappers. Brush all with egg and then place about a heaping teaspoon of the lobster filling in the middle of half of the wontons. Next, cover with another wonton and seal by wetting your fingertips with water, rubbing the edges of the wontons and pressing them together. In the process, make sure to press out any air bubbles. Repeat until you have used all the wontons. You will have extra filling most likely, which is great smeared on crostini. When ready to eat, simply boil the ravioli in salted water until ready. Serve with champagne sauce or in lobster butter.

Champagne Sauce

Ingredients
2 tbs. shallots
1 stick plus 2 tbs butter
½ cup champagne (or white wine if that's all you have)
2 tbs. heavy cream
salt and pepper to taste
chives

Directions
In a saucepan, sauté the shallots in 2 tablespoons of butter. Then add the champagne and the cream and allow it to reduce over low heat for about 7 minutes. It will lose volume, which is what you want, but be careful to not over reduce. Then cut up your very cold stick of butter into pats and grab a whisk. Turn down heat to very low. Slowly add one piece of butter at a time until all the butter has been incorporated. Turn off the heat. I know it takes a while, but it is worth it, I promise. Now taste and adjust seasoning with more salt and pepper if you like. Then add the chives. If you are serving with the

ravioli, add them now, toss gently and serve. Don't forget to drink the rest of that champagne!

Lobster Butter

Compliments of Natalie Gingerich.

Ingredients
2 lbs. unsalted high-quality butter, softened
shells of 2 lobsters

Directions
Place both ingredients into the bowl of a stand mixer with the paddle attachment. Turn to medium or high and whip for at least thirty minutes. I realize that this sounds like a ridiculous amount of time, but that is what gets all of the lobster goodness into the butter. Once the butter is a nice orange color, scrape the mixture into a heavy sauce pan and heat on low until the butter becomes liquid while watching it. You do not want to separate the butter. Then pour through a fine mesh sieve. Allow the butter to cool and harden and then shape into a log in plastic wrap and store in the fridge or freezer. This makes a great sauce.

Lobster Stew

Compliments of Chris Millar.

Ingredients
two small lobsters
$\frac{1}{2}$ cup butter
1 quart whole milk

Directions
Cook the lobsters and remove the meat immediately, saving the tomalley and blood (the white substance) from inside the shell. Simmer the tomalley and blood in $\frac{1}{2}$ cup of butter for eight minutes. Incorporate the meat (cut into big chunks), cooking slowly on a low heat for 10 minutes.

Remove from heat and let it cool slightly. Then, very slowly, add 1 quart of whole milk a trickle at a time. While adding the milk, you must stir constantly, otherwise the milk will curdle. Keep stirring until the stew blossoms into a rich salmon color. Allow the stew to stand, refrigerated, for 5 to 6 hours before reheating and serving.

Lobster Potpie

Compliments of Natalie Gingerich.

Ingredients
1 lb. bacon cut into small pieces
2 cleaned and chopped leeks
1 large handful chanterelle mushrooms (or any beautiful wild mushroom)
1 stick butter
1 onion, finely chopped
1 carrot, finely chopped
1 rib of celery, finely chopped
3 garlic cloves, finely chopped
¼ cup flour
¼ cup white wine
2 tbs. brandy
2 cups milk (or 1 cup lobster stock and 1 cup milk)
salt and pepper
2 cups lobster meat
3 ears of corn cut off the cob
½ cup chopped chives
2 homemade or store-bought piecrust shells
1 package puff pastry
1 egg

Directions
Cook bacon until brown and crisp and then remove from the fat and drain. Pour most of the fat out and sauté the leeks and chanterelles in the same pan. Allow to cool. To make the sauce, melt the butter in a saucepan and add the onion, carrot, celery and garlic and sauté for a few minutes. Then add the flour and stir well for a few minutes. Add the wine and brandy and cook for a few minutes, then add the milk. Season with salt and pepper to

taste. The sauce should be very thick. This is how I like it. However, if you like it looser, feel free to add more liquid. With a thick sauce, the end result is a slice of potpie.

In a mixing bowl, combine the leek/mushroom mixture with the chopped lobster meat, corn and chives. Divide between the pie shells. Sprinkle bacon over and top with the sauce, and then cover with the puff pastry. Use a fork to seal the puff pastry to the pie dough. Make a hole in the center of the pastry for steam to escape and decorate the top with any trimmings from the pastry. Brush entire pie with egg wash and bake at 450 degrees until golden brown. Enjoy!

LOBSTER GRILLED CHEESE SANDWICH

Ingredients
meat from one lobster
Gruyere cheese
butter
bread slices

Directions
Simply cook your lobster and pick out the meat. Break the meat into chunks and then sandwich it between thin layers of grated Gruyere cheese on your favorite buttered bread. Toast on a George Foreman grill (or in a pan) until the cheese is melted. A sprinkle of cayenne pepper would give it some extra edge!

A TRADITIONAL CLAMBAKE (LOBSTER BAKE)

Of all the great American cookouts, surely the lobster bake, known outside of Maine as the clambake, is the most historic and dramatic. The clambake tradition is actually older than America itself, learned by the Pilgrims from Native Americans. According to historical lore, the Pilgrims watched Native Americans gather lobster, clams, fish, corn and potatoes and prepare them on the beach. The Native Americans dug a sand pit and lined it with hot rocks and coals. They added the lobsters, clams, fish, corn and potatoes to the pit and covered them in fresh, wet seaweed and more hot rocks, steaming the lobster and shellfish in seawater.[85]

Nowadays, the clambake is a popular New England sunset dinner. The menu and cooking method can vary a bit. Below is one version of the classic clambake.

Ingredients
3 to 4 pounds round clams
6 to 10 large baking potatoes
6 medium onions, peeled
6 to 10 ears of corn (husk left on but silk removed)
12 live lobsters
12 lemons cut into wedges
lots of melted butter
enough seaweed to cover the fire pit
a tarp (tarpaulin)
plenty of cheesecloth

Directions
Start by digging a hole in the ground, covering the bottom with large stones and building a big fire on top of the stones. The fire should burn for at least two hours to get the stones good and hot. In the meantime, you can prepare the food. Wrap individual servings of the above ingredients in cheesecloth, tying the corners together. Once the rocks are hot enough to spit a drop of water back at you, rake off the coals from the fire and cover the rocks with seaweed. Place the food packets on the seaweed and cover with more seaweed. Cover the entire hole of food with a wet tarp, sealing the steam created by the hot stone and seaweed (allowing a very small amount of steam to escape to relieve the pressure). After about 2 hours or once the potatoes are soft, everything should be done. Serve the bake with melted butter. Don't forget napkins—you'll need them!

NOTES

In order to expand the contents of this book beyond my personal views and experiences, I developed and mailed a questionnaire to two hundred Down East Maine lobster fishermen in September 2011. All of these fishermen were members of the Down East Lobstermen's Association (DELA).

In the questionnaire, I asked thirty questions, including but not limited to how long the participant had been lobstering, whether or not he was from a generation of fishermen, whether or not his children were involved in lobstering, if he'd sustained any injuries while lobstering and how satisfied he was with lobster fishing as his job (on a scale of 1 to 10, with 10 being really satisfied). I also asked a variety of open-ended questions, including what he liked most and least about being a fisherman, the biggest change he'd seen in the industry since starting fishing and what most concerned him about the future of Maine lobster fishing.

I received replies from sixty of the fishermen. While this is not a statistically significant sample given the total number of Down East lobster fishermen, I found the responses very enlightening and worth including in this book. For example, I was pleasantly surprised to see that 82 percent of respondents rated their level of job satisfaction at 8 or above. I was also amazed to see how consistent fishermen were on some of the open-ended questions. For example, when asked to write what they love most about being a lobster fisherman, 32 percent spontaneously replied "being my own boss," another 30 percent answered "independence" and a further 14 percent replied "freedom." Again, I thank all the fishermen and women who took the time to fill out and mail back my questionnaire so I could include their perspectives in this book.

Introduction

1. Madeleine Hall-Arber, Christopher Dyer and John Poggie, "New England's Fishing Communities," MIT Sea Grant, 2001, 386, seagrant. mit.edu/pubs_desc.php?media_ID=11&search=new+england+fishing+ communities&qt=all&fa=&categ=.
2. Data from the Maine Department of Marine Resources.
3. Ibid.; Hall-Arber, Dyer aand Poggie, "New England's Fishing Communities," 380.
4. James M. Acheson, *The Lobster Gangs of Maine* (Hanover, NH: University Press of New England, 1988), 2.
5. Colin Woodard, *The Lobster Coast* (New York: Viking, 2004), 266.
6. Author's research; Gulf of Maine Research Institute, "Taking the Pulse of the Lobster Industry: A Socioeconomic Survey of New England Lobster Fishermen," 2008, 28, www.gmri.org/upload/files/gmri_lobster_report_ lores.pdf.
7. Hall-Arber, Dyer and Poggie, "New England's Fishing Communities," 386.
8. OECD Insights, "A Lesson in Resources Management from Elinor Ostrom," oecdinsights.org/2011/07/01/a-lesson-in-resources-management-from-elinor-ostrom.
9. Woodard, *Lobster Coast*, 268.

A Brief History of Lobster Fishing in Maine

10. The Maine Department of Marine Resources, "The History of Lobstering," www.maine.gov/dmr/rm/lobster/guide/index.htm#history.
11. Acheson, *Lobster Gangs of Maine*, 4.
12. Gulf of Maine Research Institute, "Taking the Pulse of the Lobster Industry."
13. Acheson, *Lobster Gangs of Maine*, 4; Woodard, *Lobster Coast*, 176.
14. Woodard, *Lobster Coast*, 175–79; Lobster Institute, "History," www. lobster.um.maine.edu/index.php?page=52.
15. Acheson, *Lobster Gangs of Maine*, 5.
16. Lobster Institute, "History."
17. Trevor Corson, *The Secret Life of Lobsters* (New York: HarperCollins, 2007), 118; Lobster Institute, "History."
18. Woodard, *Lobster Coast*, 181–85.

19. Lobster Institute, "History"; Virginia L. Thorndike, *Maine Lobster Boats: Builders and Lobstermen Speak of Their Craft* (Camden, ME: Down East Books, 1998), 19.

20. Data from the Maine Department of Marine Resources; Maine Lobster Promotion Council; R.S. Steneck, T.P. Hughes, J.E. Cinner, W.N. Adger, S.N. Arnold, F. Berkes, S.A. Boudreau, K. Brown, C. Folke, L. Gunderson, P. Olsson, M. Scheffer, E. Stephenson, B. Walker, J. Wilson, B. Worm. "Creation of a Gilded Trap by the High Economic Value of the Maine Lobster Fishery. Conservation Biology," 2011. doi: 10.1111/j.1523-1739.2011.01717.x.

21. Data from the Maine Department of Marine Resources.

Frequently Asked Questions About Lobster Fishing

22. Lobster Institute, "Cooking & Eating Lobster," www.lobster.um.maine.edu/index.php?page=22.

The Amazing American Lobster

23. Lobster Institute, "Life of the American Lobster," www.lobsterinstitute.org/index.php?page=20.

24. Corson, *Secret Life of Lobsters*, 128–29.

25. Lobster Institute, "Life of the American Lobster."

26. Corson, *Secret Life of Lobsters*, 117, 130.

27. How Stuff Works, "Is There a 400-Pound Lobster Out There?" science.howstuffworks.com/environmental/life/zoology/marine-life/400-pound-lobster.htm.

How to Catch a Lobster

28. Grundéns, "Grundéns History," grundens.com/index.php?option=com_content&view=article&id=72&Itemid=172.

29. Gulf of Maine Research Institute, "Taking the Pulse of the Lobster Industry."

30. Data from the Maine Department of Marine Resources.

THE STRATEGY OF LOBSTER FISHING

31. Acheson, *Lobster Gangs of Maine*, 21.
32. Ibid., 91.
33. Ibid., 98; author interview with Norbert Lemieux.
34. Hall-Arber, Dyer and Poggie, "New England's Fishing Communities," 388.
35. Christopher Little, *The Rockbound Coast* (New York: Norton, 1994).
36. Author interview with Lemieux.

THE SEASONS OF LOBSTER FISHING

37. The Lobster Institute, "Life Cycle & Reproduction," www.lobsterinstitute.org/index.php?page=48.
38. Ibid.
39. Ibid.
40. Author's research.

THE MAINE LOBSTER MARKET

41. Acheson, *Lobster Gangs of Maine*, 118.
42. Ibid., 145.
43. Author interview with Lemieux; Acheson, *Lobster Gangs of Maine*, 15.

THE PERILS OF LOBSTER FISHING

44. Peter Rousmaniere, "Diving into a Crushing Occupational Divide," *Risk & Insurance* (March 2005), findarticles.com/p/articles/mi_m0BJK/is_3_16/ai_n13454909.
45. "Hypothermia," www.westpacmarine.com/samples/hypothermia_chart.asp.

THE UNWRITTEN RULES OF LOBSTER FISHING

46. Author's research; Acheson, *Lobster Gangs of Maine*, 35.
47. Interview with Cutler fishermen; Acheson, *Lobster Gangs of Maine*, 74.

48. Clarke Canfield, "Lobster Wars Turn Violent in Maine," *Huffington Post*, September 4, 2009, www.huffingtonpost.com/2009/09/04/lobster-wars-turn-violent_n_277748.html; *Bangor Daily News*, "Lobster Wars Still Simmering," July 20, 2010, bangordailynews.com/2010/07/20/news/lobster-wars-still-simmering.

49. Maine Supreme Judicial Court, *State v. David McMahan*, November 13, 2000, www.courts.state.me.us/opinions_orders/opinions/documents/00me200m.htm.

50. Canfield, "Lobster Wars Turn Violent in Maine"; *Bangor Daily News*, "Lobster Wars Still Simmering."

The Written Rules of Lobster Fishing

51. Author's research.

52. Corson, *Secret Life of Lobsters*.

53. Lobster Institute, "History."

54. James Acheson and Roy Gardner, "The Evolution of the Maine Lobster V-Notch Practice: Cooperation in a Prisoner's Dilemma Game," *Ecology and Society* 16 (2011), www.ecologyandsociety.org/vol16/iss1/art41; Corson, *Secret Life of Lobsters*, 121.

55. Author interview with Lemieux.

56. Corson, *Secret Life of Lobsters*, 121.

57. Maine Lobster Council.

58. The Maine Department of Marine Resources, "Maine Lobster Zone Councils," www.maine.gov/dmr/council/lobsterzonecouncils/index.htm.

How to Become a Lobster Fisherman

59. Walter Day's quote is from Carey Goldberg, "Down East, the Lobster Hauls Are Up Big," *New York Times*, May 31, 2001.

60. In reference to William Ernest Henley's poem, "Invictus."

61. Acheson, *Lobster Gangs of Maine*, 39, 42–44.

62. The Maine Department of Marine Resources, www.google.co.uk/url?sa=t&rct=j&q=&esrc=s&source=web&cd=1&ved=0CCcQFjAA&url=http%3A%2F%2Fwww.maine.gov%2Fdmr%2Frm%2Flobster%2Fapprenticebrochure.pdf&ei=tlVwT8KoNIjE0QXZ1YGOAg&usg=AFQjCNGdZHw4ONyxLKsNAtRouTg-6OpLbA.

MAINE LOBSTER BOATS THROUGH THE AGES

63. Jack Sherwood, "The Downeast Tradition at Block Dog Boat Works," www.blackdogboatworks.com/html/the_DownEast_tradition.html.

64. *Maine Boats, Homes & Harbors*, "Evolution of the Maine Lobsterboat," May–June 2010, www.maineboats.com/online/boat-features/evolution-maine-lobsterboat.

65. Ibid.

66. Ibid.

67. Ibid.

68. Nancy Beal, "Wooden Boatbuilding in Beals: An Island Tradition Fading Fast," *Fishermen's Voice*, March 2005, www.fishermensvoice.com/archives/woodenboatbldg.html; Wikipedia, "Beals, Maine," en.wikipedia.org/wiki/Beals,_Maine; Mike Crowe, "A Century of Boats," *Fishermen's Voice*, July 2004, www.fishermensvoice.com/archives/acenturyofboats.html.

69. Crowe, "A Century of Boats."

70. Ibid.

71. Ibid.

72. *Fishermen's Voice*, "Arvid Young, 1940–2011," www.fishermensvoice.com/112011ArvidYoung.html.

73. Dickerson Boat Service Log, "What Is a Downeast Boat?" oya.com/tips/2008/07/what-is-a-DownEast-boat; Jeff Della Penna, "Driving Backwards," *Fishermen's Voice*, August 2006, www.fishermensvoice.com/archives/0806index.html.

74. Dickerson Boat Service Log, "What Is a Downeast Boat?"; Della Penna, "Driving Backwards."

IT'S NOT ALL HARD WORK

75. "Holland's Boatshop: Boatbuilders & Lobsterboat Racers," video.google.com/videoplay?docid=7737684698687399984.

76. Robert Tomsho, "In Maine, a Rivalry Boils Up on the Lobster-Boat Racing Circuit," *Wall Street Journal*, August 25, 2009, online.wsj.com/article/SB125115648215255195.html.

77. Jess Bidgood, "New England Asks: Who Has the Fairest Lobster Trap Tree of All?" December 17, 2011, www.nytimes.com/2011/12/18/us/new-england-asks-who-has-fairest-lobster-trap-tree-of-all.html.

HOW TO EAT A LOBSTER

78. American Lobster, "Frequently Asked Questions," www.parl.ns.ca/lobster/faq.htm.
79. Maine Lobster, "Lobster Facts," maine-lobster.com/lobster-facts.
80. Lobster Institute, "Cooking & Eating Lobster."
81. Ibid.
82. Ibid.
83. Maine Lobster, "Lobster Facts."
84. University of Maine Department of Food Science & Human Nutrition & the Lobster Institute, "How to Freeze Lobster at Home," www.lobster.um.maine.edu/media/How%20to%20Freeze%20Lobster%20at%20Home.pdf.

MY FAVORITE LOBSTER RECIPES

85. Derek Riches, "Clambake," About.com, bbq.about.com/od/specificdishes/a/aa060201a.htm.

ABOUT THE
AUTHOR

Christina Lemieux Oragano was born and raised in Cutler, Maine, where her family has been in the lobster industry for four generations. When she was ten years old, Christina began working summers and school breaks on her father's lobster boat and continued to do so until she graduated from college.

After college, Christina began a career in advertising, working in San Francisco and New York before moving to London, where she met and married her husband, Anthony. She and Anthony continue to live in London, where they raise their daughter, Anya. Christina's continued passion for the lobster industry is evidenced through her blogging, her experimentation with new lobster recipes and her dedication to helping her fishing family during trips back to Cutler.